GLORY, HUMILIATION, AND THE DRIVE TO WAR

From Hitler's determination to erase Germany's disgrace after World War I to Sadat's promise to undo Egypt's humiliation in 1967 to ISIS's proclamations that it would end the "emasculation" of Muslims and restore the glory of the Caliphate, a sense of political humiliation and a desire for martial glory have always been central in the drive to war. Yet although glory and humiliation are the twin engines of conflict, and together they spur individuals and nations to violence, philosophers have shown little interest in these dispositions. In this book, Nir Eisikovits offers a philosophical account of political humiliation, martial glory, and the relationship between them. Drawing on philosophy, literature, and psychology, Eisikovits argues that it is impossible to understand why people are drawn to war and how wars are justified without making sense of these two political passions and the ways in which they inflame each other.

NIR EISIKOVITS is Professor of Philosophy and Founding Director of the Applied Ethics Center at the University of Massachusetts Boston. His previous books include *Sympathizing with the Enemy: Reconciliation, Transitional Justice, Negotiation* (2010), *A Theory of Truces* (2015), and *Theorizing Transitional Justice* (2016).

"In a groundbreaking book, Nir Eisikovits probes how a sense of personal and political humiliation can drive a people to war and instill a hunger for glory. Eisikovits writes with grace and compassion as he explores history and its examples, whether his lens is focused on Pericles's funeral oration, World War I and II, China, or the current wars in Ukraine and the Middle East. This is a book we need to study, *now*."

Nancy Sherman, Georgetown University

"Eisikovits splendidly brings together political theory, conceptual investigations, and case studies. His book, by directing our attention to the dangerous dynamics of glory and humiliation, provides an original and illuminating contribution to our understanding of the question why states go to war."

Thomas Brudholm, University of Copenhagen

"Nir Eisikovits' *Glory, Humiliation, and the Drive to War* offers a brilliant analysis of key aspects of the moral psychology of war frequently overlooked: political humiliation and martial glory. He powerfully argues that we will not understand why wars begin and continue without considering these drivers. Beautifully written and illustrated with rich examples drawn from a wide breadth of times and places, Eisikovits' book is essential reading for understanding the dynamics of modern conflicts."

Colleen Murphy, University of Illinois Urbana-Champaign

GLORY, HUMILIATION, AND THE DRIVE TO WAR

NIR EISIKOVITS

University of Massachusetts Boston

CAMBRIDGE
UNIVERSITY PRESS

CAMBRIDGE
UNIVERSITY PRESS

Shaftesbury Road, Cambridge CB2 8EA, United Kingdom

One Liberty Plaza, 20th Floor, New York, NY 10006, USA

477 Williamstown Road, Port Melbourne, VIC 3207, Australia

314–321, 3rd Floor, Plot 3, Splendor Forum, Jasola District Centre, New Delhi – 110025, India

103 Penang Road, #05–06/07, Visioncrest Commercial, Singapore 238467

Cambridge University Press is part of Cambridge University Press & Assessment, a department of the University of Cambridge.

We share the University's mission to contribute to society through the pursuit of education, learning and research at the highest international levels of excellence.

www.cambridge.org
Information on this title: www.cambridge.org/9781009012836

DOI: 10.1017/9781009026451

When citing this work, please include a reference to the DOI 10.1017/9781009026451

First published 2024
First paperback edition 2026

A catalogue record for this publication is available from the British Library

Library of Congress Cataloging-in-Publication data
NAMES: Eisikovits, Nir, author.
TITLE: Glory, humiliation, and the drive to war / Nir Eisikovits, University of Massachusetts, Boston.
DESCRIPTION: Cambridge, United Kingdom ; New York, NY: Cambridge University Press, 2024. | Includes bibliographical references and index.
IDENTIFIERS: LCCN 2024022009 (print) | LCCN 2024022010 (ebook) | ISBN 9781316515716 (hardback) | ISBN 9781009012836 (paperback) | ISBN 9781009026451 (epub)
SUBJECTS: LCSH: War (Philosophy) | War–Psychological aspects. | War–Causes. | Glory. | Humiliation–Political aspects.
CLASSIFICATION: LCC U21.2 .E45 2024 (print) | LCC U21.2 (ebook) | DDC 355.0201–dc23/eng/20240617
LC record available at https://lccn.loc.gov/2024022009
LC ebook record available at https://lccn.loc.gov/2024022010

ISBN 978-1-316-51571-6 Hardback
ISBN 978-1-009-01283-6 Paperback

For Masha

War is the strong life; it is life in extremis.

William James, "The Moral Equivalent of War"

Both boys dead? but that's out of nature. We all
Have been patriots, yet each house must always keep one.
'Twere imbecile, hewing out roads to a wall;
And, when Italy's made, for what end is it done
If we have not a son?

Elizabeth Barrett Browning, "Mother and Poet"

Contents

Acknowledgments

Early versions of the argument made here were presented at West Point and to the faculty colloquium of the Conflict Resolution program at the University of Massachusetts Boston. While writing, I received invaluable feedback from many colleagues and friends. I am particularly grateful to Maria Granik, Greg Fried, Jeremy Wanderer, Chris Zurn, Kenneth Greenberg, Fred Marchant, Dan Feldman, Steven Levine, Chad Finsterwald, Vladimir Petrovic, Jean-Philippe Belleau, and Alan Zebek. Some of them read earlier drafts, others commented on arguments I tried out with them. I learned a lot from what they told me.

I'm also indebted to the students in my seminars on the philosophy of war at the university over the last few years for the opportunity to discuss the ideas that animate this book. I have greatly enjoyed reading the classical texts that inform this work – from Homer to Clausewitz – with them.

I'm grateful to Hilary Gaskin of Cambridge University Press for her enthusiastic embrace of this project and for her expert guidance. Her encouragement made this work both more philosophical and wider ranging than I thought possible. Sections of some of the chapters were published in other venues. Small parts of the arguments offered in Chapters 5 and 6 appeared in my article "Proportionality and Self Interest," *Human Rights Review* 11(2) (2010). A short section of Chapter 4 appeared in "Embarrassment and Political Repair," in Thomas Brudholm and Johannes Lang (eds.), *Emotions and Mass Atrocity: Philosophical and Theoretical Explorations* (Cambridge: Cambridge University Press, 2018). Another brief section of Chapter 4 appeared in Jacob Burley and Nir Eisikovits, "Workplace Automation and Political Replacement: A Valid Analogy?" *AI and Ethics* 3(4) (2022). Parts of Chapters 4, 5, and 8 appeared in "Political Humiliation and the Sense of Replacement," in Graham Parsons and Mark Wilson (eds.), *How to End a War* (Cambridge: Cambridge University Press, 2023).

I'm grateful to my in-laws, Tanya Nosik and Sasha Granik, for their warmth and generosity over the years, and to my parents, Zvi and Rivka Eisikovits, who have been a constant source of love and support, who patiently read numerous versions of these arguments (and showed them to their friends), and have always insisted that everything I do is great. It's not true, but it's very nice to hear. My daughters Sasha, Nina, and Naomi are wonders – they bring no end of delight into my life. Spending time with them is the best antidote to the dark materials I think and write about.

There are a lot of references to poetry in this book, and I hope they shed some light on the moral psychology of war. But sometimes it's best to keep things simple. I'll stick with Nat King Cole. "The greatest thing you'll ever learn," he sings, "is just to love and be loved in return." I dedicate this book to my wife, Masha, the best person I know, who taught me the greatest thing I'll ever learn.

Introduction

In the summer of 1932, Albert Einstein wrote Sigmund Freud a famous letter, asking what it would take for humans to stop fighting each other, whether "there [was] any way of delivering mankind from the menace of war." As far as the famous physicist was concerned, the fact that wars continued unabated in spite of widespread recognition of their horror meant that the issue was, at bottom, psychological: "Man has within him a lust for hatred and destruction. In normal times this passion exists in a latent state, it emerges only in unusual circumstances; but it is a comparatively easy task to call it into play and raise it to the power of a collective psychosis." And so, the question of controlling war is really about what is possible psychologically: "Is it possible to control man's mental evolution so as to make him proof against the psychosis of hate and destructiveness?"

In his response, Freud agreed that war is made possible, in part, by the existence of destructive instincts that comingle with our other, more "erotic" ones. But our responsiveness to war is not just a question of instincts. It also turns on our preoccupation with ideas, and it is determined by the importance we attach to material interests. War has psychological, philosophical, and economical roots; it is multi-determined: "[W]hen a nation is summoned to engage in war, a whole gamut of human motives may respond to this appeal–high and low motives ... The lust for aggression and destruction is certainly included ... The stimulation of these destructive impulses by appeals to idealism and the erotic instinct naturally facilitate their release."

The force and persistence of the destructive instinct will not allow for the abolishment of war, Freud wrote. The challenge is not eliminating war but channeling our aggressive tendencies, themselves irreducible, towards less deadly outputs – a process Freud elsewhere calls "sublimation." On a slightly more optimistic note, Freud adds that we may try to strengthen ties of identification between human beings, and we may even try to socially engineer a class of intellectual leaders less susceptible to fighting.

("[A] superior class of independent thinkers, unamenable to intimidation and fervent in the quest of truth, whose role it would be to question the ways in which war is urged upon the public.") And yet, these are both very long-term projects. So long-term, they "conjure up an ugly picture of mills that grind so slowly that, before the flour is ready, men are dead of hunger."[1]

Freud was always ambivalent about Einstein. It didn't help that a few years earlier the physicist refused to recommend him for the Nobel Prize. Einstein wasn't convinced about the value of psychoanalysis: "[N]otwithstanding my admiration for the ingenious achievements of Freud, I hesitate to intervene in this case. I couldn't convince myself of the validity of Freud's theory and I am therefore unable to form an authoritative judgment for others."[2] The great psychologist, for his part, didn't think much of the younger man's psychological acumen: "[Einstein] is cheerful, confident and kind, understands as much about psychology as I do about physics, and so we had a very good conversation,"[3] he reported to a student about their first meeting in 1927. And it is indeed easy to dismiss Einstein's initial question as the earnest, naïve musing of a scientist whose grasp on history and political psychology is tenuous. But Einstein's query does, in fact, raise the most beguiling question of political philosophy: Why do people fight? Why is organized conflict one of the most constant forces – across eras and cultures – in human history? If war is so bad, why do we keep going to war?

This book attempts a partial answer. Freud is of course right that war is multi-determined. In these pages, I will focus on the moral feelings that fuel fighting. In particular, I will explore the role that political humiliation and the thirst for glory play in the run-up to and maintenance of conflict. The book offers a philosophical account of these feelings and the relationship between them, as well as a discussion of real-world cases in which they propel and animate war. Consider the following examples, a few of which I take up in detail in these pages:

On June 21st 1940, a week after the Nazis marched into Paris, Hitler ordered the "Compiègne Wagon," the train car in which the Germans signed the 1918 armistice, removed from the French museum in which it was kept. The wagon was taken back to the exact spot in which the defeated Germans signed the

[1] The entire correspondence is available here: www.public.asu.edu/~jmlynch/273/documents/FreudEinstein.pdf.

[2] www.haaretz.com/life/books/2017-05-25/ty-article-magazine/.premium/when-freud-and-einstein-tried-to-answer-an-age-old-question-why-war/0000017f-f755-d044-adff-f7fda5230000#:~:text=I%20couldn%27t%20convince%20myself,an%20authoritative%20judgment%20for%20others.

[3] Ibid.

agreement that ended the Great War. Hitler then marched a French delegation into the train car and dictated his terms for accepting their surrender. With exacting symmetry, the German humiliation in World War I was reversed.[4]

Writing in the July 2014 issue of Dabiq, the Islamic State's promotional magazine, the group's spokesperson, Abu Mohammed Al Adnani, declared: "The time has come for those generations that were drowning in oceans of disgrace, being nursed on the milk of humiliation, and being ruled by the vilest of all people, after their long slumber in the darkness of neglect—the time has come for them to rise. The time has come for the ummah of Muhammad . . . to wake up from its sleep, remove the garments of dishonor, and shake off the dust of humiliation and disgrace, for the era of lamenting and moaning has gone, and the dawn of honor has emerged anew."[5]

In an undated letter from an American federal prison, documented by the Anti-Defamation League, Jimmy Matchette, a member of the Hammerskin neo-Nazi gang, wrote to his colleagues that "Being a Hammerskin is the distinct feeling of being set apart from the entire planet. And of knowing we will conquer & overcome all obstacles to achieve our goals and accomplish our great work, knowing that if we fail, all is lost forever and the west will perish. Even though I am locked down in a maximum-security federal penitentiary, I wouldn't [have] traded the opportunity for all the gold in the world. You, my true comrades, hold all the glory of victory at your fingertips. We really are the most notorious white power Skinhead group in the entire World."[6]

A few weeks after the end of the 1967 war, Shmuel Gonen (also known as Gorodish), then commander of the IDF's 7th armored brigade, gave a speech to mark his group's victory in the battles for control of the Sinai Peninsula: "[E]verywhere we passed through we left behind burnt skeletons, destroyed tanks, and charred bodies," he told his soldiers. "We looked straight at death, and death lowered its eyes . . . My warrior comrades! You did not know the majesty of your own courage, and when I saw the armored vehicles burning and pierced and the people- you- fighting on from within, I came to know that man was steel and the armor mere metal . . . in this war victory is all yours. Be strong and brave, my brothers, heroes of glory!"[7]

4 www.iwm.org.uk/history/what-you-need-to-know-about-the-battle-of-france.

5 Full copies of the magazine have been removed from the Internet since conducting the research for this book. But the quote is available here: https://eaworldview.com/2014/06/iraq-text-isis-declaration-caliphate-muslims/.

6 www.adl.org/education/resources/profiles/hammerskin-nation.

7 http://yadlashiryon.gal-ed.co.il/Info/hi_show.aspx?id=51236&t=3 (my translation from Hebrew).

Writing in his 1977 autobiography *In Search of Identity*, Egyptian President Anwar el Sadat explained the rationale for his country's surprise attack against Israel in 1973: "[T]he basic task was to wipe out the disgrace and humiliation that followed from the 1967 defeat. I reckoned it would be 1000 times more honorable for us . . . to be buried crossing the [Suez] Canal than to accept disgrace and humiliation."[8]

From Hitler's determination to erase Germany's disgrace in Versailles to Sadat's promise to undo Egypt's humiliation in 1967 to ISIS's proclamations that it would end the "emasculation" of the *ummah* and restore the glory of the Caliphate, a sense of political humiliation and a desire for martial glory have always been central in the drive to war. But though glory and humiliation are in many ways the two-stroke engine of conflict, and though, taken together, they spur individuals and nations to violence, contemporary political philosophers have shown little interest in these dispositions. There have been few philosophical works on political humiliation and how it drives belligerence, and even fewer on the meaning of glory. In fact, with one notable exception, the last time political philosophy took glory seriously was when Hobbes wrote about it in the *Leviathan*.

In this book, I offer a philosophical account of political humiliation, martial glory, and the relationship between them. I argue that it is impossible to understand why people are drawn to war and how wars are justified without making sense of these two political passions and the ways in which they inflame each other. In Part I, I offer a philosophical account of martial glory. I argue that glory still matters, even in the age of asymmetrical war and drone strikes, and that part of our ambivalence about the term comes from a failure to distinguish between personal and political glory. In Part II, I suggest that political humiliation is more than the flip side of recognition and that it often consists in the sense of lost status and anxieties about replacement. I describe how the sense of humiliation travels across time and space (making us feel humiliated because of events that happened long ago or because of what befell people we had never met), and then I develop a theory of humiliation in foreign policy. In Part III, I explore how political humiliation and martial glory combine in the lead-up to war and take up three case studies.

More specifically, Chapter 1 offers a philosophical history of glory beginning with Homer and Thucydides, through the works of Cicero, Machiavelli, and Hobbes, and concluding with a discussion of the meaning of the term for Hannah Arendt. Chapter 2 draws on these historical

[8] Anwar Sadat, *In Search of Identity: An Autobiography* (New York: HarperCollins, 1978): p. 215.

materials to construct a theory of glory that distinguishes between personal and political glory, examines the relationship between glory and honor, considers the scope of the concept (what kind of achievements can be deemed glorious?), and investigates its relationship to luck and the passage of time. The theory also distinguishes between glory as a motivation for acting and the bestowal of glory as an ex post facto act of social recognition. The chapter then goes on to discuss the gap between promises of glory and who actually gets glorified, the relationship between glory and death, and some normative questions including the relationship between glory and duty and the degree to which the bestowal of glory follows the precepts of just war theory. Chapter 3 tries to explain the modern ambivalence about glory. It goes through a series of "arguments" against glory, including the historical claim that the culture of honor breeds violence and should, therefore, be replaced by an emphasis on moneymaking, the argument that emerged after World War I that glory no longer makes sense in the wake of mechanized warfare, the incompatibility of glory with the practices of combatants in asymmetrical wars, the incompatibility of glory and the rise of technologies facilitating killing at a distance, and the comparative significance of glory and camaraderie for the soldiers who actually fight wars. At the end of the chapter, I suggest that much of the modern skepticism about martial glory is a result of a tendency to run together two distinct understandings of the term – namely personal glory (of the kind that Achilles sought in the *Iliad*) and political glory (of the kind that Pericles promised in his funeral oration).

Chapter 4 begins to sketch a theory of political humiliation by examining Avishai Margalit's recognition-based theory of humiliation. I argue that this theory, while important, does not consider the possibility that there is more to humiliation than the denial of recognition and that it also fails to take seriously key aspects of the phenomenology of humiliation, such as the fact that it can be contagious and, more importantly, the sense of being replaced that lies at the heart of feeling humiliated. The chapter concludes by drawing clear distinctions between humiliation, shame, and embarrassment, and by considering the impact of technology on the role humiliation plays in our lives. Chapter 5 develops this theory of political humiliation further by offering a philosophical account of humiliation in international relations. A humiliating foreign policy involves acting vis-a-vis others simply because one can count on their inability to respond, disregarding the national heritage or history of other nations, dismissing and ignoring the complex social and cultural realities making up the lives of those we come into contact with, failing to take another group seriously as a subject of promises, and viewing another political entity primarily as a source of resources. I explain the relationship between these elements and provide historical examples that illustrate how they play out in the real world. The chapter also develops further the argument made in Chapter 4, that the sense of being replaced is central to political humiliation, and suggests several reasons why it is important to understand the role of replacement in contemporary politics. The chapter continues with a consideration of the special role of humiliation in autocratic regimes and concludes with a discussion of the dual role of humiliation in international affairs – as both a spurn to war and a method of fighting. Chapter 6 argues that the need to avoid humiliation is already present in key areas of political theory. In particular, it informs precepts of *jus in bello* requirements in just war theory, serves as a rationale justifying various transitional justice policies, and helps make sense of our thinking about appeasement. Chapter 7 considers the relationship between glory and humiliation. It argues that glory is most often bestowed for ending a history of political humiliation and that humiliation is often understood in terms of lost glory. In this chapter, I also argue that there is a lopsidedness in the social distribution of humiliation and glory. While women (and children and the elderly) are the primary recipients of wars' humiliations, men tend to enjoy most of the glory. Finally, Chapter 8 offers three case studies for illustrating the theory advanced in this book: the 1973 Middle East war, the rise of ISIS, and Russia's revanchist foreign policy in the last fifteen years.

As I write this Introduction, the Kremlin's war of aggression in Ukraine, justified by claims of Russia's historical humiliation after the collapse of the Soviet Union, is well into its second year. Henry Kissinger, celebrating his 100th birthday, continues to vigorously advocate a "great powers" view of international relations, in which the interests of smaller nations are dismissed, even though this tendency to humiliate minor actors has sparked wars from the Peloponnesus to the "Levant" and the Balkans over the centuries. In much of the developed world, a postindustrial malaise has resulted in widespread feelings of hopelessness and humiliation that have fueled a frightening rise of right-wing populism, a trend that threatens democratic institutions. And these humiliations – real and manufactured – are always attended by promises of great honor and glory for anyone fighting to undo them. While philosophers have been looking elsewhere, the combustible mix of glory and humiliation continues to shape the political world. This book aims to bring political philosophy up to speed in this crucial area.

PART I

Glory

CHAPTER 1

A Short History of Glory

Our idea of political and military glory is incongruous. On the one hand, its legitimacy and even its coherence have been called into question by the rise of mechanized and unmanned warfare, and by the ubiquity of asymmetrical fighting. Wilfred Owen's complaint that it's grotesque to lionize the glory of patriotic death after the industrial slaughter in the battlefields of the Great War[1] has become something of a cliché about modern fighting. On the other hand, military glory is not going anywhere: From Marine Corps recruitment videos to the throngs mourning Iranian General Qasem Soleimani, from ISIS war poems extolling the restoration of the Caliphate to alt-right talking points offering praise and great renown to those who would keep America white and Christian, the rhetoric of military glory is still as central to organized violence as it was for Homer in the eighth century BCE. How are we to make sense of this tension? Before we make headway on that question we need to understand more clearly what we mean when we speak about glory. This chapter will offer an overview of thinking about glory, and this history will, in turn, supply resources for the theory of glory offered in the next chapter. Only with the theory in place can we understand the modern ambivalence about martial glory. What follows is not a comprehensive intellectual history. I'm not an expert on the work of the thinkers discussed in this chapter. I offer, instead, a philosophical mining expedition aimed at extracting the raw materials needed for a historically informed theory of martial glory. As this chapter will make clear, with some exceptions, it has been centuries since philosophers have taken glory seriously. Given that the term is very much present in political discourse and practice, that is a troubling lacuna.

The idea of military glory is as old as war (or at least as old as writing about war): Glory or *kleos* figures prominently as a theme in Homer's *Iliad*

[1] See Wilfred Owen, "Dulce et Decorum Est," in Cecil Day-Lewis (ed.), *The Collected Poems of Wilfred Owen* (New York: New Directions Books, 1963).

and Thucydides' *History of the Peloponnesian War* – the two key ancient Greek texts about war.[2] Cultural and literary references to glory abound in other great works – from Shakespeare's rendering of King Henry's speech before the Battle of Agincourt ("And gentlemen in England now a-bed shall think themselves accurs'd they were not here"[3]) to Victor Frankenstein's admonition of Captain Walton's mutinous crew ("Did you not call this a glorious expedition? And wherefore was it glorious? Not because the way was smooth and placid as a southern sea, but because it was full of dangers and terror"[4]). But while the term shows up repeatedly in literary and historical accounts of war and high adventure, it has had a more checkered life in political philosophy. Glory figures prominently for the Romans (particularly for Cicero) and in the thought of early modern political philosophers (mainly Machiavelli and Hobbes), but with the exception of Hannah Arendt (who addresses the idea somewhat indirectly), there has been little recent conceptual work on the subject.[5] Let's take a closer look.

1.1 Homeric Glory

The natural starting point for any historical account of glory in the Western tradition is Homer's *Iliad*. The narrative contours of the epic are familiar: Achilles is part of the coalition that travels to Troy under King Agamemnon. The expedition is occasioned by the abduction of Helen – the King's sister-in-law. Like most of the Achaeans, Achilles does not believe in or care about the justness of the coalition's cause, nor does he believe in the competence of its leaders. He fights not for king or country but because he was *made* to fight and die a glorious death. Indeed, when Agamemnon takes possession of Achilles' war captive, Briseis, Achilles complains to his goddess mother, Thetis, that he is being denied the opportunity to achieve the glory and honor that are the very premise of his existence: "Since, my mother, you bore me to be a man with a short

[2] For an interesting overview about the role of *kleos* – primarily in the *Odyssey* but with some references to the *Iliad*, see Charles Segal, "Kleos and its Ironies in the Odyssey," *L'antiquité Classique* 52 (1983): 22–47.

[3] William Shakespeare, *Henry V*, Act 4, Scene 3. http://shakespeare.mit.edu/henryv/henryv.4.3.html.

[4] Mary Shelly, *Frankenstein* (London, UK: Collectors Library, 2017): p. 256.

[5] A good deal of the contemporary literature on glory takes up what Hobbes and Machiavelli had to say about the term and we shall look at some key interpretations in this chapter. A notable and rare contemporary account of glory is offered in Timothy Chappell, "Glory as an Ethical Idea," *Philosophical Investigations* 34(2) (2011): 105–34. We will have more to say about the Chappell essay later.

life, therefore Zeus of the loud thunder on Olympus should grant me honor at least."[6] War for Achilles is not judged by its justice or necessity; for him, it is not, in a phrase Clausewitz would make famous later, an instrument of policy. Rather, it is "where men seek glory"[7]: an opportunity to display excellence and achieve renown.

There is no celebration of war in the *Iliad*. If anything, the heartbreaking vignettes it contains about the lives of the Greeks and Trojans who were killed in combat prefigure the "pity of war" themes in World War I poetry.[8] There is no elevation of a grand cause worth dying for. The war that the Achaeans fight is not a good war. And yet Achilles floats above all of this. When he fights, he is in his element, someone doing what he was meant to do – to be excellent at violence. On this account, war is a testing ground, a place for "proving one's mettle." Skill, danger, and courage are all required for passing that test. The kind of glory at work here is mostly *amoral*. What is at issue is the ability to perform well – exceptionally well – under conditions that would undo most. The ability to do with élan what most are afraid to do at all. As Dr. Johnson points out, "the profession of soldiers and sailors has the dignity of danger. Mankind reverence those who have got over fear, which is so general a weakness."[9] Of course, Achilles is ambivalent about the value of martial glory. Homer portrays him as keenly aware of the alternative to dying young – namely returning home and living out his life peacefully. Indeed, when Odysseus finds Achilles in Hades in Book XI of the *Odyssey*, the warrior's ghost dismisses his glorious death, remarking ruefully, "by god, I'd rather slave on earth for another man – some dirt-poor tenant farmer who scrapes to keep alive – than rule down here over all the breathless dead."[10]

Achilles' counterpart in the *Iliad*, the Trojan prince Hector, is also not invested in the cause of the war. As far as he's concerned, his brother, Paris, recklessly entangled the city in an unnecessary fight. And yet Hector cannot but fight. The silliness of the cause does not grant him a reprieve. The glory is in fighting well, in leading the Trojan soldiers (as his position demands), and, more than anything, in avoiding shame or any perception

6 Homer, *Iliad*, trans. Fagles (New York: Penguin, 1988): book 1, line 352.
7 Ibid., book 6, line 160.
8 Owen used this phrase, which later came to epitomize the human cost of the Great War, in the introduction he planned, before his death, for a poetry collection. He wrote: "My subject is War and the pity of War. The Poetry is in the Pity." See Owen, *Collected Poems*, p. 31. For an interesting discussion of Homer's capacity for sympathy for both sides of the Trojan War, see Caroline Alexander, *The War that Killed Achilles* (New York: Penguin, 2010).
9 James Boswell, *Life of Johnson*, ed. Chapman (New York: Oxford University Press, 1980): p. 927.
10 Homer, *The Odyssey*, trans. Fagles (New York: Penguin, 1997): book 11, line 557.

of cowardice. Like Achilles, Hector is keenly aware of the potential costs of the war. In one of the epic's most moving exchanges, Hector's wife, Andromache, begs him to wait, to fight defensively. She has nobody left in the world but him: "You, Hector – you are my father now, my noble mother, a brother too, and you are my husband, young and warm and strong."[11] Hector knows that if the Trojans lose, the war would claim the lives of his parents and siblings. And even that, he tells his wife, would be "nothing, nothing beside your agony."[12] He knows she might be whisked off to the land of Argos, "laboring at a loom, at another woman's beck and call."[13] And yet fight he must, for he would "die of shame"[14] if he avoids battle. His role is "to stand up bravely always to fight in the front ranks of the Trojan soldiers, winning my father great glory, glory to myself."[15] For Hector the thirst for glory mixes with a keen sense of duty and a fear of acting shamefully. The proportions between the elements of this mixture are probably a bit different for each of the two heroes. Achilles is motivated more by personal glory, Hector perhaps more by a sense of duty; we shall have more to say later about the relationship between the duty to fight and the bestowal of glory. But, strikingly, in both cases and throughout the *Iliad*, martial glory is completely disconnected from the reasons for going to war in the first place.

1.2 Thucydides on Glory

If Achilles' idea of glory is fundamentally apolitical, focused on the recognition of great skill, Thucydides offers us a diametrically opposed view. Consider Pericles' famous funeral oration, made a year or so into the Peloponnesian War. The beginning of the oration is awkward. Pericles complains that he is uncomfortable giving a speech to commemorate Athens' fallen soldiers because such an address makes the reputation of many turn on the rhetorical skill of one (himself). The relatives of the dead are bound to be disappointed by any speech (as no praise can satisfy them), while those who did not know the soldiers will suspect hyperbole. After this uncomfortable introduction, Pericles goes on to deliver what has become the most celebrated eulogy in history.

The address begins with a glowing description of Athens – its democratic institutions, the leisure and high standard of living it affords, and the tolerance and respect for individual pursuits – all of which make for a

[11] Ibid., book 6, lines 508–10. [12] Ibid., line 539. [13] Ibid., line 543. [14] Ibid., line 523.
[15] Ibid., lines 527–29.

unique way of life worth fighting and dying for. Athens is glorious because its institutions and its citizens' way of life are glorious. And those who die to defend these institutions and this way of life thus deserve and receive glory. Their glory, according to Pericles, consists in an erasure of their past misdeeds, elevating their lives from the mix of achievement, mediocrity, and foul behavior that marks our own. Dying for a cause as noble as Athens lifts the dead out of drab ordinariness; it erases their personal imperfections and makes their names live on forever: "[E]ven those who were inferior in other ways deserve to have their faults overshadowed by their courageous deaths in war for the sake of their country."[16] With his sacrifice, "each man won praise for himself that will never grow old."[17] Those who have died for a noble cause go down in history, Pericles says, and are remembered not only during the lifetime of those whom they helped or in the places they fought for, but always and anywhere – "for to famous men all the earth is a monument."[18]

Glory for Pericles, then, is the robust recognition and honor that come from the willingness to sacrifice for a great cause. In its purest form, it is bestowed on the dead whose faults it erases and whose reputation it makes everlasting. Pericles notes that the dead receive this full-blooded version of glory because people have an easier time honoring those who are no longer in their way: "[Our] good will towards the dead is free of rivalry."[19] Those who were willing to sacrifice their lives for a great cause but survived will likely receive a significant degree of glory but it will, in the nature of things, be somewhat diminished: "[E]ven extreme virtue will scarcely win you a reputation equal to [that of the dead]."[20] There is, then, according to this view, an interesting relationship between glory and dying. First, the willingness to die for a great cause is part of what one is glorified for. Second, the act of actually dying for the cause facilitates the bestowal of glory by the living: Since those who are left behind are the ones who ultimately award glory, it is easier for them to lavish praise on those who no longer compete with them. Finally, this account tells us that getting oneself killed for an important political project is a useful tool for reputation management. A glorious death revamps one's social standing in a way that does not quite apply to those who stay alive, subject, as they are, to the ravages of time and their own future behavior.

[16] Thucydides, *On Justice, Power and Glory: Selections from the History of the Peloponnesian War*, trans. Woodruff (Indianapolis, IN: Hackett Publishing, 1993): book II, section 42.
[17] Ibid., section 43. [18] Ibid. [19] Ibid., section 45. [20] Ibid.

It's interesting to note that Pericles' military strategy early in the war with Sparta – calling on his compatriots to abandon their farms, crowd in behind the walls of Athens, refuse direct engagement with the Spartans, and rely on their navy to both resupply the city and harass its enemies – was original, innovative, even daring. But it was not glorious – at least not in the Achillean sense. The bravado of engaging in combat, the excitement of it, are missing from Pericles' strategy, and are replaced by a strategic brilliance that, it was hoped, would exhaust the Spartans. The glory Pericles is interested in is the glory of the city itself – preserving it was foremost in his mind. Those who would preserve it would merit glory not by virtue of how fiercely they wielded weapons but by virtue of the political project they saved.[21]

Of course, Pericles is not the only character in Thucydides' vast panorama who is interested in glory. There are certainly actors in the *History* who, closer to the Achillean mold, pursue it for self-aggrandizement rather than for political reasons. Chief among these is Alcibiades, who rashly advocates for the Sicilian Expedition in order to increase his personal renown. Alcibiades' interest in glory is narcissistic – he is what Hobbes would later call "vainglorious," insofar as he holds himself in higher esteem than he deserves. Rather than winning glory because he served the city, Alcibiades uses the city and engineers its political course in order to increase his own glory. He was, to quote Thucydides, "exceedingly ambitious of a command by which he hoped to reduce Sicily and Carthage and personally to gain in wealth and reputation by means of his success."[22]

1.3 Cicero

Cicero's *On Obligations* offers the most extensive treatment of glory in the Roman tradition. The orator apparently wrote a comprehensive treatise titled *On Glory*,[23] but that book has been lost. The discussion in *On Obligations* likely summarizes that work.

[21] On Pericles' war strategy, see Donald Kagan, *The Peloponnesian War* (New York: Penguin, 2004), particularly part 2. See also Michael Palmer, "Love of Glory and the Common Good," *American Political Science Review* 76(4) (1982): 825–36. Palmer writes that Pericles chose an "inglorious, prosaic strategy requiring the old element of Athens to abandon its sacred hearths, temples, and monuments in order to move into the city and [the] new element to suppress its daring, its inclination for motion, to remain at rest" (pp. 827–28).

[22] Thucydides, *History*, book VI, section 15.

[23] For a discussion of the lost work, see, for example, Evan Dutmer, "Summa et Perfecta Gloria: Cicero on Ambition, Reputation, and Care for Future Human Beings," *Ethical Perspectives* 29(1) (2022): 7–31.

Cicero begins by noting that glory should not be reserved for military achievements.[24] His skepticism about the status of military exploits is likely tied to the fact that he was not a military man himself, and was keen to establish that civic accomplishments – of the kind he was responsible for during his legal career and consulship – were as important as any martial triumph.[25] In fact, he tells us, military achievements can't be more important than civic ones *exactly because* the former pursuits are often animated by a desire for winning glory rather than by the substantive importance of the project. The common assumption that accomplishments in war are most worthy of recognition has to be questioned. Civic achievements, like the legislation the famous lawgiver Solon wrote for Athens, are to be ranked higher than military victories accomplished by the likes of Themistocles (the hero of the Battle of Salamis between the Greeks and Persians in 480 BCE), because while military victory "brought the state only momentary advantage . . . Solon's reform [is] . . . of perennial benefit to it."

Moreover, the promulgation of prudent policy makes military victories possible more often than the other way around – by creating the resources that support fighting, or by attracting the civilian support necessary for going to war. Therefore, we must "let arms unto the toga yield, the laurel to men's praise."[26] That Cicero's arguments for the primacy of civilian achievements are self-serving does not make them wrong. As Christopher Hitchens once put it in a moving essay on the legacy of Edmund Burke, "[T]he fact that you have located a man's most selfish motive does not mean you have identified his best motive."[27] Cicero may have been both vain and correct about the importance of civic leadership.

Even more strikingly, Cicero suggests that civic achievements deserve the kind of glory that we typically reserve for military pursuits, not only because the former feats turn out to be more beneficial but also because they require the same kind of courage: "[T]here are instances of courage, then, on the home front which are on par with bravery in war; indeed they

[24] Cicero, *On Obligations*, trans. Walsh (New York: Oxford University Press, 2008): book 1, sections 74–76.

[25] For the classic treatment of Cicero's recasting of glory away from the traditional Roman militaristic conception, see Anthony Long, "Cicero's Politics in *De Officiis*," in Anthony Long (ed.), *From Epicurus to Epictetus: Studies in Hellenistic and Roman Philosophy* (Oxford: Oxford University Press, 2006): pp. 307–34.

[26] Cicero, *On Obligations*, book I, section 77.

[27] Christopher Hitchens, "Edmund Burke: Reactionary Prophet" in *The Atlantic* (April 2004 Issue). www.theatlantic.com/magazine/archive/2004/04/reactionary-prophet/302914/.

demand more application and effort than those in war."[28] Cicero does not quite explain how it is that policymaking requires physical courage. Perhaps he means that in Rome's raucous political culture a miscalculation could put one in the same mortal danger that a military mistake would cause for a soldier. If that is what he had in mind, his claim is not wholly without merit.

Book II of *On Obligations* offers a sophisticated conceptual analysis of glory, but before diving into it, Cicero makes an interesting psychological observation: While it is always useful to be glorified or honored "both in general and in forging friendships in particular," the need for glory and honor is different among different people.[29] The first part of the statement is puzzling: How does being glorified help in forging friendships? And why would anyone want friends who were primarily interested in one's glory? But the latter comment is insightful. We do not all want glory equally – and this may be true even among members of the political class. The need for glory, which Hobbes would posit as a universal feature of human psychology, operates differentially within different people. It's likely that those who choose to live in the public eye are more drawn to glory than introverts who shun the limelight. But even among those more public figures some leaders are more retiring than others. One thinks, for example, about the difference between a Jimmy Carter and a Bill Clinton or, across the ocean, between a Shimon Peres and a Yitzhak Shamir. An interest in glory cannot be universally assumed in all political leaders.

Having made this observation, Cicero moves on to account for what goes into being glorified. He argues that the attainment of glory depends on three cumulative conditions: the affection of common people towards the exalted person, their trust in that person, and their belief (mingled with admiration) that the recipient is worthy of being glorified. And so, we learn that for Cicero glory turns on the attitudes of the nonaristocratic class. The affection of the commons (the first condition) is gained primarily by services rendered. (Did the exalted figure do nice things for "the people" like arranging handouts or putting on compelling entertainments?) It can also be obtained if the figure tried to render such services (even if unsuccessfully) or, finally, if that person has a reputation for generosity, kindness, justice, good faith, and other similar virtues. The second factor centers on the cultivation of trust. Cicero tells us that we trust those who we think are prudent (people who can deal with unexpected

[28] Cicero, *On Obligations*, book I, section 78. [29] Ibid., book II, section 31.

challenges better than we can) or those we believe to be good and just (those to whom we can hand over the care of our affairs and be reasonably confident they won't take advantage of us). Between the two, Cicero argues that being just is more important: "Justice without prudence will be highly effective but prudence without justice will make no impact."[30] As in other parts of his work, it's not completely clear if Cicero is making a descriptive or normative point here: Is he arguing that we should trust the just more than the prudent or that we in fact do? And it's far from clear whether he is right about either the normative or descriptive claims (people may well actually trust a prudent leader more than a just one if they think the former can more readily advance their interests, and if government has an obligation to be effective, perhaps that is as it should be). As for the third factor, admiration, Cicero tells us that we admire great achievements that confound our expectations. We admire those who excel others in merit and are able to resist common weaknesses which others succumb to. Thus, we admire those who will give up money or disdain danger in favor of an important cause. Cicero concludes by arguing that the just person fills all three conditions for glory – he will attain affection and trust because he seeks to benefit the greatest number of citizens and he will attain admiration because he has the character traits that most of us lack and is able to despise temptations and danger.

Cicero does not argue very closely for this purported connection between glory and justice. Notably missing are considerations of outcome – the just leader is not necessarily the most successful one. Cicero does not go deeply into the comparative role of success and character as the grounds for glory. Thus, he does not include the creation of lasting political goods or the attainment of great military victories among his criteria for attaining glory. A relatively modest political leader who keeps the state on course, is a benefactor of the people, and behaves decently meets all of Cicero's criteria for glory. As we will see shortly, this view will be roundly rejected by Machiavelli. Decency and glory may both be good, but they are not, the Florentine philosopher insists, the same thing.

Cicero moves finally to discuss glory's temporal horizon. Becoming glorious, he tells us, is a long-term project: "[I]f people imagine they can obtain enduring glory by deceit and empty exhibitionism and hypocrisy in word and look, they are wildly off the mark; true glory drops roots and also spreads its branches wide whereas all false claims swiftly wither away like frail blossoms."[31] It is here that Cicero most explicitly ties glory with

[30] Ibid., section 34. [31] Ibid., section 43.

morality. True glory is lasting and one only gets glorified for what deserves glory – for what is actually good. This is why he argues that glory "must be systematically both inquired and invested." Glory is not the same as the fleeting adulation of contemporaries – it is about acting in a way that will, in time, be recognized as worthy of the highest praise. Once more, the account is set up to highlight the glory of the statesman rather than that of the warrior. While the triumph of the latter will likely result in immediate admiration (which may fade if it turns out that the military project was not in the real interest of the state), the statesman's patient building of institutions and infrastructure can result in lasting glory, which will become manifest only after these political achievements start bearing fruit.

1.4 Machiavelli

For Machiavelli glory is the reward for great feats in domestic politics, diplomacy, and war.[32] Thus, for example, founding kingdoms and republics is a sure way to win political glory.[33] Setting up despotic cities, in contrast, cannot result in glory. Machiavelli's conception of glory has a normative component. While he famously argues that it is necessary for a prince to learn how "not to be good" and to use that skill as needed in order to maintain the state, while it may be necessary for the prince to be both lion and fox, while he might need to lie, dissemble, and practice cruelty in order to stay in power – to do these things gratuitously cannot yield glory. Knowing how to be wicked and how wicked one must be in order to stay in power and keep one's city strong is one thing – the prince who gets that balance right will win political glory and deserve it – but wickedness for its own sake is another. A murderous leader who fails to economize on cruelty cannot be described as glorious even if successful. Hence the famous discussion in *The Prince* of Agathocles – who in spite of his remarkable achievements cannot be described as glorious because he was incapable of practicing restraint.[34]

[32] Russell Price, "The Theme of Gloria in Machiavelli," *Renaissance Quarterly* 30(4) (1977): 588–631, p. 592.

[33] Ibid., p. 594.

[34] As Machiavelli puts it, "[O]ne cannot call it virtue to kill his fellow citizens, to betray his friends, to be without faith, without pity, without religion; which modes enabled him to acquire imperium but not glory." Niccolo Machiavelli, *The Prince*, trans. Alvarez (Long Grove, IL: Waveland Press, 1989): chapter 8. For an interesting dissenting view arguing that Machiavelli actually saw Agathocles as worthy of political glory, see John McCormick, "Machiavelli's Inglorious Tyrants: On Agathocles, Scipio and Unmerited Glory," *History of Political Thought* 36(1) (2015): 29–52.

Machiavelli limits the proper application of the term "glory" to exalted military or political achievements. The label is a form of "super recognition" primarily open to the cream of the political class and Machiavelli would have likely resisted widening its application to more pedestrian activities such as athletic competitions or scholarly attainment.[35] Further, for Machiavelli, glory is not a "side effect" of good governance; it's not something that simply accrues to a leader in the course of ruling well. Rather, it's something that a leader should actively pursue. Refusing to do so can lead to the decay of the state; for a prince to dismiss the pursuit of glory is a problem in itself – an indication of flagging political energy. Politics should not become the realm of patient, low-key management;[36] rather, seeking glory is what the true leader does. True, a prince can't become glorious if he becomes completely unmoored from principles, but to turn from glory, to dismiss it, to reject its pursuit as hubristic, is to practice the kind of otherworldly politics that Machiavelli disdains.[37]

So while Machiavelli, like other Renaissance thinkers, seeks to revive Greek and Roman understandings of political virtue, while he accepts the humanistic premise that worldly success is, to a significant degree, a function of agency rather than of luck, while he rejects the Christian admonition offered by Boethius, Aquinas and their ilk to renounce the quest for worldly glory because we can't control whether or not we obtain it, his interpretation of the virtues is different from that of Cicero or Seneca. The virtues necessary for winning glory go beyond the traditional, earnest list that Cicero enumerated (including wisdom, justice, honesty, courage, and temperance). They must, instead, include the ability to lie and cheat when needed. Machiavelli sets himself in direct opposition to the Romans here. In Book II of *On Obligations*, Cicero warned that "if anyone thinks that he can win lasting glory by pretense, by empty show, by hypocritical talk and looks, he is very much mistaken."[38] Machiavelli, in contrast, famously insists that these vices are sometimes necessary for ruling well. The question is not whether a prince practices them, but whether he does so parsimoniously and in the service of a great cause.

35 Price, Gloria in Machiavelli, p. 621. 36 Ibid., pp. 608–9.

37 For excellent discussions of the contrasting attitudes of humanists and "schoolmen" to worldly glory, see David Owen, "Machiavelli's *il Principe* and the Politics of Glory," *European Journal of Political Theory* 16(1) (2015), and more broadly, Quentin Skinner, *Visions of Politics Vol 2: Renaissance Virtues* (Cambridge: Cambridge University Press, 2002). Skinner quotes Aquinas whose view is paradigmatic of the schoolmen: "[T]he desire for human glory destroys any magnanimity of character ... the duty of a good man is to show contempt for glory and all such temporal goods" (p. 122).

38 Cicero, *On Obligations*, book II, section 43.

Political feats often require pretense, empty show, and hypocritical talk. These do not disqualify anyone from becoming glorious. Machiavelli, the great nonideal theorist, learned that much from his missions on behalf of the Florentine Signoria and Dieci to the courts of King Louis, Cesare Borgia, and Pope Julius II. The virtues of the prince will include what the classical humanists considered vices. Achieving glory is going to require knowing how to practice these vices prudently, knowing how "not to be good," and deploying that knowledge selectively.[39] The main question is not whether the prince will lie and pretend – for lie and pretend he must to maintain his rule and achieve political greatness. Rather, the question is whether he lies *more* than he must, whether he pretends gratuitously.[40]

While Machiavelli's focus is on the prince's pursuit of glory, he does recognize that the term can motivate the people as well. Many years in Florence's diplomatic service taught him how vulnerable a state was if it depended on mercenary armies for defense. Mercenaries were notoriously fickle, and because they fought for money, they would often change sides if offered a better deal. The Florentines were not, for example, served well by their mercenary forces in their attempts to keep control of Pisa. Both as a policymaker and as a theorist, Machiavelli argued for the superiority of militia armies, and in 1509, he organized and successfully led a Florentine militia against Pisa. Militiamen fought not for money but for the security of the state and the interests of their kin. Their soldiers were, in other words, animated by patriotism, and for such fighters, there could be glory in the fight.

In chapter 25 of *The Prince*, as he reflects on the conduct of Pope Julius II, Machiavelli offers us an interesting observation about the relationship between glory and luck. Julius was impetuous and full of energy; he liked to charge forward, to *act* rather than carefully consider possible courses of action. Those characteristics happened to work for the geopolitical circumstances of the time – Julius was, as we are fond of saying, the right man at the right time. And, as Machiavelli notes dryly, he died early enough that the times did not change and make his character an impediment: "I wish to let stand his other actions, which were all the same, and all of which succeeded well for him. And the brevity of his life did not let him feel the contrary. For if times would have come which would have made him need

[39] Machiavelli, *The Prince*, chapters 15, 18.

[40] For a detailed discussion, see Quentin Skinner, *Machiavelli: A Very Short Introduction* (New York: Oxford University Press, 2001): chapter 2.

to proceed with care, his ruin would have followed – he would have never deviated from those modes to which nature inclined him."[41]

In other words, much of Pope Julius' legacy, much of whatever glory he obtained, was a result of luck, or what contemporary philosophers sometimes call "moral luck" – he benefitted from a fortunate congruity between his psychological makeup and the political moment he operated in – an overlap between "constitutional luck" and "circumstance luck."[42] Had he lived longer, he would have probably met with failure as conditions changed and called for a person of greater prudence. Glory, this discussion suggests, at least when it is understood as a lasting legacy, is intertwined with luck; it depends on the circumstances that prevailed when a leader was in power, on how long they lived, and on how (a factor Machiavelli does not consider) social conventions evolved during and after the leader's lifetime. Churchill lived long enough for his character to both unmake him because of his impetuosity, and remake him by virtue of his courage and pugnaciousness when the Nazis came along. And, as social mores evolved after his death, Churchill's staunch imperialism and his attitude towards women had, with time, dampened his legacy again.[43]

Finally, in the *Discourses on Livy*, Machiavelli also offers us an interesting meditation on the risks of glory for those who stay alive long enough to claim their own renown. If Pericles provides an argument about why it is easier to glorify the dead, Machiavelli gives us a reminder of how dangerous glory can become for the living. A military commander who returns gloriously triumphant from battle has typically earned so much respect and reverence from fellow soldiers, enemies, and citizens that he poses an immediate threat to the political leadership of the country that sent him. Glory will often generate a toxic ingratitude towards those who won it and lived to tell: "[T]he commander who has skillfully acquired an empire for his lord by overcoming his enemies and showering himself with glory, and his troops with riches, necessarily acquires among his own soldiers and his enemies and among the subjects of his prince such renown that his victory cannot prove other than distasteful to the very lord who sent him forth."[44]

41 Machiavelli, *The Prince*, chapter 25.

42 See Thomas Nagel, "Moral Luck" and Bernard Williams, "Moral Luck," both reprinted, in Daniel Statman (ed.), *Moral Luck* (Albany, NY: SUNY Press, 1993).

43 For a discussion on shifting legacies in the context of the debate over controversial monuments, see Nir Eisikovits, "Not Set in Stone: Five Bad Arguments for Letting Monuments Stand," *Journal of Global Ethics* 16(3) (2020): 404–13.

44 Niccolo Machiavelli, *Discourses on Livy*, trans. Bondanella (New York: Oxford University Press, 2008): book 1, chapter 29.

1.5 Hobbes

Our preoccupation with glory is, according to Hobbes, an important reason why we can't live without a strong government to keep us in check. That fixation is a psychological feature that both makes us dangerous to each other in the absence of government, and that gives the government, once established, something powerful to work with when it wants to motivate us to pursue certain political projects.

Our interest in glory can be retrospective or forward-looking. Retrospective glory consists in the pleasure one takes in superior past achievements. A forward-looking desire for glory drives the seeker to obtain more power than others (so that she can, later on, *glory in* these attainments). As Gabriella Slomp notes, the centrality of glory for Hobbes recedes as we move to his later texts.[45] Whereas in earlier writings, a desire for winning more glory than others is the main driver of human conflict, generating competition and jealousies, by the time he writes the *Leviathan*, Hobbes comes to see glory as one of several passions (in addition to fear and the desire for scarce resources) that generate strife. Even in the later work, though, glory and honor seeking are important for understanding why humans, unlike, say, ants and bees, can't live sociably without government: "[M]en are continually in competition for Honour and Dignity, which these [other] creatures are not; and consequently, amongst men there ariseth on that ground, Envy and Hatred, and finally Warre; but amongst these not so."[46]

Hobbes famously offers a distinction between glory and vainglory. The former is grounded in a factually correct reading of one's history of achievements and how they are perceived by others. The latter consists in taking empty flattery too seriously and developing an inflated sense of one's worth. The vainglorious are legends in their own minds: They think they achieved more than they actually have, or they are happy to listen to those who falsely tell them that's the case. Vainglory is a passion that can lead to criminality as the vainglorious often assume they are above the law. Vainglory also makes those afflicted more prone to anger since their inflated sense of self leads them to feel insulted easily.[47]

45 See Gabriella Slomp, "From 'Genus' to 'Species': The Unravelling of Hobbesian Glory," *History of Political Thought* 19(4) (1998): 552–69, and, in far greater detail, Gabriella Slomp, *Thomas Hobbes and the Political Philosophy of Glory* (London, UK: Palgrave MacMillan, 2000).

46 Thomas Hobbes, *Leviathan*, ed. and introd. Curley (Indianapolis, IN: Hackett Publishing, 1994): 17.7.

47 Ibid., 6.39, 27.13, 27.16–17.

While the traditional reading of Hobbes by scholars such as Strauss, Walzer, and Pettit emphasizes the competition for recognition as an important cause of strife, others, like Clifton Mark, focus on the destabilizing role of insult.[48] On this interpretation, the same desire for glory that prompts citizens to jostle for comparative position also makes them "prickly"[49] or sensitive to offense. Such prickliness can incite violence over trivialities – "for trifles, as a word, a smile, a different opinion, and any other sign of undervalue, either direct in their persons or by reflection in their kindred, their friends, their nation, their profession, or their name."[50]

Some interpreters disagree about how psychologically central the desire for glory really is for Hobbes. According to Slomp, Hobbes attributes greater motivational force to self-preservation than to glory. The passions, according to her reading, will rarely lead us to die rather than endure humiliation.[51] On Peg Birmingham's reading, in contrast, we want to be admired so badly that we become dangerous when we can't have the recognition we crave; thus, it is the desire for glory, more than anything else, that generates the suspicion and fear which, in turn, necessitate the state.[52]

Regardless of its psychological gravity, it should be stressed that glory, for Hobbes, is a social passion; its full realization depends on the institutions of civil society. We certainly desire glory in the state of nature, but given the relative equality between agents in that state, it doesn't make sense to view oneself as consistently superior to others or to demand recognition of our superiority. As Hobbes put it, "[T]he question who is the better man has no place in the condition of meer nature."[53] The law, by defining standards of value and worth, introduces the possibility of actually winning a competition based on one's talents and efforts as well as the security necessary to enjoy the results of that victory.[54]

Given Hobbes' moral skepticism, the objects of glory can't have an objective status. It's a psychological fact about us that we want glory, as far as Hobbes is concerned, and given that psychological fact, it's a social fact

48 For a discussion of both views, see Clifton Mark, "The Natural Laws of Good Manners: Hobbes, Glory, and Early Modern Civility," *The Review of Politics* 80(3) (2018): 391–414. For a fascinating account of how this tendency to "prickliness" shaped the culture of honor in the antebellum American South, see Kenneth Greenberg, *Honor and Slavery* (Princeton, NJ: Princeton University Press, 1997).

49 Clifton Mark borrows the term from Arash Abizadeh. See Mark, Good Manners, p. 402.

50 Hobbes, *Leviathan*, chapter 13.

51 Slomp, Hobbesian Glory, p. 558.

52 Peg Birmingham, "Arendt and Hobbes: Glory, Sacrificial Violence, and the Political Imagination," *Research in Phenomenology* 41(1) (2011): 1–22, p. 9.

53 Hobbes, *Leviathan*, chapter 15.

54 Slomp, Hobbesian Glory, p. 562.

that governments must designate objects of glory. But the objects so designated are not glorious in themselves; the goals that the state marks as glorious – as important enough to endow those who pursue them with glory – are made such by the very act of designation, by political decision. Just as with his beliefs about morality or the status of laws, for Hobbes the objects of glory are glorious by convention rather than by nature.[55]

1.6 Arendt

In a 1944 article for *Aufbau*, a periodical published in New York and aimed primarily at German Jewish emigres, Arendt approvingly quotes the leaders of the Warsaw Ghetto Uprising who described the rebellion as an act that salvages "the honor and glory of the Jewish people."[56]

The rebels, Arendt wrote, stepped beyond an ethos of victimhood. Doing that "ended the pariah existence of the Jewish people in Europe" and positioned them as equal actors in the struggle against the Nazis: "[B]y claiming equal rights [Jews] joined the ranks of other European peoples in the struggle for freedom." Political equality and glory are acquired, on this view, by the willingness to become master of one's own fate, take a risk, and die for a cause. The cause was primarily the liberation of Jews: "[T]hey intend to battle as Jews for the freedom of Jews." The shape of the new Europe, Arendt thought, was being molded by underground movements, and in fighting the Nazis in Warsaw, Jews claimed their place in that effort. And, because they have, one could now start plausibly speaking of Jewish honor and glory: "Honor and glory are new words in the vocabulary of our people," she wrote, excitedly. "We would perhaps have to go back to the days of the Maccabees to hear such language." To be glorious, then, on this early reflection from Arendt, was to be willing to take a mortal risk for a grand project, and the project, in this case, was the establishment of an active Jewish identity that would earn its place among other national movements.[57] The theory of glory on offer here is essentially the same as the one articulated by Pericles in the funeral oration. Glory was

[55] For a discussion of Hobbes' conventionalism, see, for example, Richard Tuck, *Hobbes: A Very Short Introduction* (New York: Oxford University Press, 1989): chapter 2.

[56] The article was published on April 21, 1944. It was titled "For the Honor and Glory of the Jewish People" and is reprinted in Hannah Arendt, *The Jewish Writings*, eds. Kohn and Feldman (New York: Schocken Books, 2007): p. 199.

[57] For a useful discussion of Arendt's early conception of glory and her eventual repudiation of it, see Birmingham, Arendt and Hobbes.

the result of sacrificing for a worthy cause; the renown is refracted from the political purpose served.

Arendt's later work, especially *The Human Condition*, contains a more nuanced discussion of glory. Humans, she argues, are characterized by the refusal to be determined by the constraints of the "human condition" (mortality and confinement to the planet to name two). As she puts it, "[O]ur life conditions never fully condition us."[58] If glory's reward is a kind of immortality or escape from our mortal confinement, as Pericles suggests, Arendt provides us with a context for understanding why such a reward is so appealing. Humans are the only mortal species, Arendt argues, since only they possess individual self-awareness including the realization that one day they will cease to exist. It is this awareness of impending individual extinction that makes people interested in immortality in the first place. And for Arendt, the most promising route to immortality is through lasting works: "[B]y their capacity for the immortal deed, by their ability to leave non-perishable traces behind, men, their individual mortality not-withstanding, attain an immortality of their own and prove themselves to be of a 'divine' nature."[59] Part of what makes glory so attractive to us, in other words, is that we *desperately* want the immortality it promises, simply by virtue of being human.

In chapter 2 of *The Human Condition*, Arendt sketches a "Greek" understanding of politics. According to this view, politics is a second life away from the privacy of home. If "home" is concerned with meeting life's physical demands, politics is an ascent from such necessities and is, thus, the realm of freedom.[60] Politics is the realm of words and actions divorced from the "merely" useful. Far more being a mechanism for problem-solving, politics is the context where free and thoughtful beings can be most themselves.[61] For Arendt, then, the political is not all about finding an efficient way to adjudicate private concerns and protect property, as it would be in a Lockean view. Property is important but only as the instrumental condition for facilitating interactions that are not about property or its administration.[62] The teleology of politics as expressed in the liberal tradition is thus reversed: Private property gives us a place to hide from the public when we need to rest, but that private space is what makes possible an ascent to politics.[63]

[58] Hannah Arendt, *The Human Condition* (Chicago, IL: The University of Chicago Press, 1989): p. 11.
[59] Ibid., p. 19. [60] Ibid., pp. 24–25. [61] Ibid., p. 26. [62] Ibid., pp. 26–27, 61.
[63] Ibid., pp. 37, 70, 71, 72.

The modern state, Arendt argues, has become the realm of collective housekeeping.[64] Society is the public organization of the means of survival, it is merely the household writ large[65] and that is a perversion of what politics means or at least of what it once did for the Greeks. Necessity is a pre-political phenomenon. Meeting our basic needs can't be the main object of politics.

In her juxtaposition between Greek and modern politics, Arendt gives us a diagnosis, an explanation, for why the interest in glory has subsided in the last few centuries. Glory is about immortality, about ambitions aimed beyond the limitations of the everyday. But contemporary politics, according to her, is consumed with the everyday and its management. The modern political realm has turned into a "social" realm and the social realm understands people in statistical terms. Politics is now the realm of social science and political economy and is focused primarily on predicting and regulating large-scale behavior.[66] The social is concerned with standardization and understanding the "behavior" of large groups rather than with the "deeds" of excellent, free, and equal agents. Glory, of course, depends on deeds – on individual courage, excellence, and other virtues. Consider this, based on Auden's "The Unknown Citizen,"[67] which captures Arendt's idea of the social, with its quantitative and statistical preoccupations, perfectly:

> Our researchers into Public Opinion are content
> That he held the proper opinions for the time of year;
> When there was peace, he was for peace: when there was war, he went.
> He was married and added five children to the population,
> Which our Eugenist says was the right number for a parent of his generation.
> And our teachers report that he never interfered with their education.
> Was he free? Was he happy? The question is absurd:
> Had anything been wrong, we should certainly have heard.

Small wonder that Auden admired Arendt's work as much as he did. "Every now and then, I come across a book which gives me the impression of having been especially written for me," he said of *The Human Condition*.[68]

On the Greek view, the public realm offers engagement with what is immortal and permanent – ideas, institutions, great works. Whatever

[64] Ibid., pp. 28–29. [65] Ibid., pp. 44, 46, 60, 69. [66] Ibid., pp. 40–41.
[67] Wystan Hugh Auden, *Collected Poems* (New York: Vintage, 1991): p. 252.
[68] Auden's endorsement on the back cover of Arendt's *The Human Condition*; see Arendt, *The Human Condition*.

glimpses of immortality humans can get, they get in their political activities.[69] The modern reimagining of politics as an instrument for the regulation of private lives bespeaks a decline of interest in this type of immortality. Creating permanent works, leaving behind a lasting legacy, becoming known for great deeds – these ambitions are no longer integral to the practice of politics. In so far as glory is tied up with a desire for permanence and immortality as Thucydides suggests, this change in the nature of politics is also a move away from glory.

The Greeks had a contempt for what Arendt calls "laboring" – the effort required to sustain our physical needs and survival. This disdain stemmed from "a passionate striving for freedom from necessity and a no less passionate impatience with every effort that left no trace, no monument, no great work worthy of remembrance."[70] On their view, humans are fully alive only when they strive to leave something behind that outlasts them. Simply providing for the means of subsistence and coordinating the basic needs of many individuals would have struck them as anti-political, as an organized activity that ignores the potential of our species. "All ancient estimates of human activities," Arendt wrote, "… rest on the conviction that the labor of our body which is necessitated by its needs is slavish."[71] We now live in a laboring and consuming society. The key political project in such a society is to make the objects of consumption abundantly available: "[W]e have almost succeeded in leveling all human activities to the common denominator of securing the necessities of life and providing for their abundance."[72] This elevation of "labor" in modern political practice and thought creates societies that are less violent than before. In antiquity, violence was used to subjugate some in order to free others from labors. But, according to Arendt, the people now living under these relatively peaceful conditions have become subjugated instead to an ethos of laboring and consumption.[73]

Later in *The Human Condition*, Arendt gives us her famous account of action (as distinct from labor and work) and that discussion sheds further light on what glory does and does not mean for her. Our actions and speech are directed at others and take place in the presence of others (be that presence actual or imagined). Thus, the meaning of any action is not exhausted by what the actor intended by it. Each act is projected into a web of existing human relationships and connections (and, in turn, makes that web more intricate and richer). Since this is the nature of action – that

[69] Arendt, *The Human Condition*, pp. 55, 56. [70] Ibid., p. 81. [71] Ibid., p. 83.
[72] Ibid., p. 126. [73] Ibid., p. 129.

it always enters into a complicated relational web where others are acting as well – we are never fully in control of the consequences of our actions.[74] This understanding of action suggests that, for Arendt, the glorious deed is not a case of the individual molding the environment with their determination and courage. Rather, a deed becomes glorious through its reverberations in our relational web – through responses to it and interpretations of it. Action largely gains its significance ex post facto. Like some of the other thinkers we have surveyed, Arendt is offering us a comment on the uncertainty of glory: Though we are interested in it because we seek permanence, judgments about anyone's glory depend on a history of perceptions, receptions, and interpretations. Actions will become glorious as their significance ricochets around this relational web, and the legacy they leave is not guaranteed to be everlasting. Churchill's reputation may be up in the aftermath of Chamberlain's appeasement of Hitler and Britain's stance in the first year of the war, it may then decline after the war, decline even more when his unapologetic imperialist sympathies become unacceptable to a growing swath of people, and then it may rebound, decades later, when a new set of tyrants is on the march and people think back fondly on the old leader's courage in the face of foreign bullying. There's no blueprint to comprehensively evaluate an action in advance in the same way you can evaluate an artifact's design. Because action takes shape in a web of relationships, its meaning is both boundless and unpredictable, and its full significance is gradually articulated and rearticulated in time.[75]

In her posthumously published "Introduction into Politics," Arendt shares some further observations about the conditions for ascribing political glory. Reflecting on contemporary war, she argues that nuclear weapons cannot be understood as a means for conducting politics in the Clausewitzian mode – nuclear war is not the continuation of politics by other means – because they destroy the very conditions for politics, namely the "world" or the relational web between people.[76] There can be no glory in a nuclear war because such a war undermines the possibility of action (including glorious action), by destroying the context, the set of connections into which we act. Under these circumstances, glory could, perhaps, be reserved for those who avert wars – in the process literally saving the world. One thinks here, for example, of Khrushchev and Kennedy managing to defuse the Cuban Missile Crisis. In other words, Arendt's argument suggests that, when changes in military technology threaten the

[74] Ibid., pp. 90, 183, 184. [75] Ibid., p. 193. [76] Ibid., pp. 156–57, 162.

very possibility of action, glory, rather than being found in military exploits, will be found in avoiding them. If this extrapolation is plausible, it brings her position closer to that of Cicero than to that of Machiavelli.

Later in that same essay, Arendt tells us that the locus of greatness differs between the Greeks and Romans. For the former, politics happens inside the polis, in the agora – between equal and free citizens who ascend from the privacy of their homes. International affairs, in contrast, are simply the realm of power – the strong doing what they can and the weak accepting what they must, to quote Thucydides. Greatness is found, as Pericles argued in the funeral oration, in the city's institutions, the way of life it affords, and its culture. Glory is allotted to those who excel inside that form of life and, arguably, to those who defend it when a defense is needed.[77] For the Romans, however, politics is primarily about expanding permanent alliances; glory adheres to foreign policy projects. If for the Greeks the law creates a boundary around the polis in which greatness can be pursued, for the Romans the very ability to extend boundaries, to make legal arrangements and contracts that grow the empire, is the locus of glory.[78]

1.7 Summary

In this chapter, we offered a whirlwind tour of the history of thinking about glory. We began with the "Achillean" conception of the term, which is focused on celebrating how rather than why one fights; such glory is predicated on technical skill and achievement rather than the kind of project for which one fights. We then contrasted this idea with its "Periclean" counterpart, wherein glory is fundamentally moral and political, a form of recognition that is based not on one's aptitude but on their willingness to take a risk for a noble cause. Next, we discussed Cicero's classical account of glory. The Roman orator argues that civic pursuits are more worthy of glory than military ones, both because the former often make the latter possible and because they frequently are more closely aligned with the state's true interests. Cicero also holds that true glory unfolds in time and that there is an inherent connection between personal virtue, the capacity for doing justice, and deserving high recognition. Machiavelli is far more circumspect about the connection between personal virtue and glory. While he argues that glory has a normative dimension, it is completely possible to glorify a prince who neglected the traditional virtues in order to maintain the state. For Machiavelli, an

[77] Ibid., pp. 164–65, 178–79. [78] Ibid., pp. 180–82.

interest in glory was constitutive of competent leadership and the objects of glory were necessarily exalted: success in war, high diplomacy, or institution building on a grand scale. Hobbes' emphasis is more psychological – our need for glory, he claims, both makes us dangerous enough to each other to require the social mediation offered by government, and gives governments a tool they can use to direct individuals towards their purposes. Finally, we considered the connection Arendt drew between a "Greek" understanding of politics, where the private realm is subordinated to public "action," and the emphasis on immortality and permanence fundamental to the idea of glory.

With this overview of thought about glory in mind, we now proceed, in Chapter 2, to explore some of the key conceptual questions the pursuit of glory raises and to construct a historically informed, rudimentary theory of political glory.

CHAPTER 2

A Theory of Glory

Let's build on the historical account offered in Chapter 1 to construct a rudimentary theory of glory. That theory should help us answer some of the questions we will ask later about why we are so ambivalent about glory, why theorists have lost interest in the term, and whether it is still important to philosophically engage with it.

2.1 Personal vs. Political Glory

One key distinction that emerges from the discussion in Chapter 1 concerns personal and political glory: The former alludes to the exploits of Achilles in the *Iliad*, whereas the latter is based on what Pericles (as reported by Thucydides) had to say in his famous funeral oration, a year into the Peloponnesian War between Athens and Sparta. Personal glory is awarded to those who display exceptional skill or ability in a given activity. For Homer the context for that display is war. Achilles cared not at all for why the Greeks sailed to Troy; the moral heft of their cause was immaterial to him. War was where men pursued glory, they found it by being better than others at fighting, and that is what he intended to do. This Achillean ethos emphasizes physical ability, courage, skill, and professionalism, not the political context of the war. Echoes of this approach survive in the attitudes of contemporary military families. The scions of distinguished Prussian, British, and American military families are bound not only to defending a particular political order but also (sometimes primarily) to the tradition of military professionalism that distinguished their ancestors.[1]

[1] Thus, for example, a 2012 press release from the Laughlin Airforce Base in Texas recounts the family story of 2nd Lieutenant Keely Yankie who had just earned her flying wings. Yankee's great-great-grandfather fought in the Civil War and her grandfather in World War II. During her graduation ceremony, the infantry sword which belonged to the former was presented to her. The newly minted pilot recounted: "I was speechless and breathless when I received the sword," adding, "I was thinking

The amoral tinge that characterized Achilles is still detectable (though perhaps not as decisively) in these cases as well.

Political glory, in contrast, is based not on skill but on a willingness to take risks and (if needed) sacrifice one's life for a valuable cause. It was this kind of glory that would accrue to the defenders of Athens, according to Pericles, and it was this kind of glory that Lincoln promised the dead at Gettysburg. The conceptual boundary between personal and political glory is porous. People animated primarily by personal glory may obviously still care about a political cause, and those who fight mainly to defend such a cause may do so with more or less skill. In short, one can certainly be the subject of both kinds of glory (consider the soldier who is willing to die for the cause and displays an unusual level of military acumen – perhaps the master bomb technician or the particularly skilled special operator). But the differences in emphasis are still discernible. To summarize: On the Achillean account, glory is the recognition that comes from fighting well – it is awarded for the combination of physical prowess, skill, and the ability to face down danger. Achillean glory is amoral. It celebrates the ability to do what not many can do well, namely charge into rather than away from physical harm and function well under great duress. Periclean glory is fundamentally moral or at least fundamentally teleological – it is awarded for defending a worthy cause. It focuses not on a fighter's skill in doing so but on their willingness to sacrifice, and it is that willingness that bestows the rewards.

2.2 But What is Glory? And How is it Related to Honor?

The distinction between personal and political glory, though central, does not yet give us a working definition of glory. What, then, is glory? Glory is a particularly elevated form of honor, esteem, or regard, that is, a supercharged form of recognition. One can be honored, esteemed, or highly regarded without being glorious but not the other way around. Glory is distinguished from honor, esteem, and regard primarily by degree – it is both more exclusive (fewer people receive it) and longer lasting. Glory is attended by a promise of immortality. Those who win it receive praise that "will never grow old." Glory (or at least its political variety) also revamps

about how lucky I was to get the sword, and how it was not about me. It was about the 150 years of history and how this sword ties all of the family history together." Remarkably for an official air force press release, there is precious little about patriotism in the piece. A major part of Keely's motivation to serve clearly has to do with living up to her family's traditions. www.laughlin.af.mil/News/Article/354741/military-heritage-forged-in-steel/.

and elevates one's reputation – erasing their misdeeds and marginalizing their weaknesses. It turns the debtor, the absent parent, the questionable spouse, into a hero. This idea of glory as a form of recognition involving immortality and elevation of reputation is key to Pericles' speech, as we discussed in Chapter 1. But it also makes an appearance in other accounts of glory – literary and historical. Take, for example, Shakespeare's *Henry V*: "[T]his story shall a good man teach his son," says King Henry to his men before the Battle of Agincourt in the bard's imagining, "and Crispin Crispian shall ne'er go by from this day to the ending of the world, but we in it shall be remembered, we few, we happy few, we band of brothers."[2] Those who fought that day by Henry's side would establish ties of kinship with the king, for "he today that sheds his blood with me shall be my brother be he ne'er so vile this day shall gentle his condition."[3] In Shakespeare as in Pericles, glory comes with promises of eternal life and a social upgrade. The coupling of glory with immortality is also clearly reflected in Lincoln's Gettysburg Address: "The world will little note, nor long remember what we say here," he insisted, "but it can never forget what they did here."[4]

Let us take a closer look at the relationship between glory and honor. Philosophers have not been careful to distinguish the two terms. Some of the classical accounts of glory – such as those put forward by Cicero and Hobbes – use the terms interchangeably. Contemporary philosophers have neglected the question of honor until quite recently and have not really discussed the relationship between honor and glory. And yet, honor is making a bit of a philosophical comeback recently, both because theorists have started realizing that people seem to actually care about it even when they (the theorists) don't, and because philosophers are beginning to see ways in which the disposition can be harnessed for good.[5]

So let us pin down a few features of glory by thinking about some of the features of honor. In his 2010 book *The Honor Code*,[6] Anthony Appiah defines being honored as deserving respect. That definition, in turn, is derived from Frank Stewart's influential view of honor as having a right to

[2] William Shakespeare, *Henry V* (New York: Signet Classics, 1998), Act 4, Scene 3. [3] Ibid.

[4] The full text is available here: https://presidentlincoln.illinois.gov/visit/whats-inside/exhibits/online-exhibits/gettysburg-address-everett-copy/.

[5] For a good overview of the recent philosophical scholarship on honor, see Dan Demetriou, "Honour," in *Routledge Encyclopedia of Philosophy* (Taylor and Francis, 2020). www.rep.routledge.com/articles/thematic/honour/v-2. doi.org/10.4324/9780415249126-L036-2.

[6] Anthony Appiah, *The Honor Code* (New York: Norton, 2010).

respect.[7] If we are correct to understand glory as an amplified form of honor, then, on this account, to be glorious is to be deserving of a great deal of respect.

For Stewart and Appiah, there are two kinds of honor: *horizontal honor* (or, in Appiah's term, "recognition") that stems from one's status – as an English gentleman, a combatant, a member of the clergy, and *vertical honor* (for Appiah, "esteem") which tracks one's comparative achievements inside a status group.[8] Thus, a soldier would deserve horizontal honor for being a member of the armed forces and vertical honor for their comparative skills in marksmanship. How does this distinction inform our discussion of glory? What we have called personal or Achillean glory is vertical: It recognizes an individual (and sometimes a group) for exceeding their peers in combat-related skills. Political or Periclean glory, in contrast, is more complicated: On the one hand, it glorifies people based not on their particular skill but on their willingness to take a risk or die for a cause. In that sense, Periclean glory is a form of horizontal honor. In the context of Pericles' funeral oration, if you were one of the soldiers who were willing to die for Athens, you were, by definition, glorious. Comparative skill was not involved. But, on the other hand, if the relevant reference group for Periclean glory is the entire Athenian population, if Periclean glory is all about elevating certain people by virtue of their willingness to fight, then Periclean or political glory is also a form of vertical glory.

In *The Honor Code*, Appiah describes a dynamic wherein the objects of honor change: Foot binding and dueling, for example, are initially viewed as honorable practices by the Chinese elites and British aristocracy, respectively, but the social judgment of these practices shifted as they gradually became sources of embarrassment for those supporting them. Appiah's account of moral revolutions turns on understanding how such shifts occur. But the centrality of honor in supporting any social practice is, for him, a constant. The Chinese custom of foot binding died out when it became clear to the relevant elites that the practice made them a laughing stock abroad and, thus, hurt their honor. The locus of honor then slowly shifted away from maiming women. A similar dynamic explained the decline of dueling and the waning of support for the slave trade in Britain.[9] While honoring abides, Appiah seems to say, the objects of honor change. Is the same true for glory?

[7] Frank Steward, *Honor* (Chicago, IL: The University of Chicago Press, 1994).

[8] See Appiah, *The Honor Code*, chapter 1.

[9] For a discussion on the decline of dueling, see Appiah, *The Honor Code*, chapter 1. For a discussion of foot binding, see Appiah, *The Honor Code*, chapter 2.

To some extent, this depends on which practices are the appropriate objects of glory. If glorying is primarily reserved for military exploits, the disposition will not display the flexibility that honoring displays. If, however, the objects of glory can shift in the same way the objects of honor have, glorying would follow a similar dynamic to the one Appiah charts. As a descriptive matter, we are still glorifying war and those who are good at it, much in the same way the ancient Greeks did. This has two possible explanations: Either we have not yet had the moral revolution that would displace war as an object of glory (war, in other words, has not yet gone the way of foot binding or dueling), or the objects of glory are more constant than the objects of honor. If the second option is true, glory is not only an extreme case of honor, but also a rigid category of honoring reserved (primarily) for military exploits. In which case, if the importance we attach to such exploits declines, the practice of glorying will wither with it rather than shift to a different activity.

Finally, Appiah argues that one's own honor is shaped by the honor afforded or denied to the group they belong to. This fact is crucial in how moral revolutions come about. If enough people feel that a current practice (for example, honor killings, feet binding) discredits their collective reputation, the practice will die out. And vice versa – the honoring of a group's practices can have cascading benefits to individuals associated with that group (for example, English abolitionism reflected well on the English).[10] The second part of these dynamics is applicable to glory: Political glory afforded to a class of individuals will reflect well on many other individuals affiliated with that class (we think well of Ukrainians because of the great bravery of their armed forces; Ukrainian noncombatants can – and they deserve to – glory in the achievements of their soldiers).

2.3 The Scope of Glory

Another important conceptual question concerns the proper scope of glory. Does this kind of recognition apply primarily to great military deeds or attainments in international affairs, or can it also cover achievements in domestic policy and other areas such as sports and entertainment? As we saw in Chapter 1, for Machiavelli, glory was reserved for the former domain: He understood it as a form of recognition that primarily denoted martial triumphs and great achievements in statecraft. Machiavelli would have probably not objected to glorifying achievements in domestic policy

[10] Appiah, *The Honor Code*, chapters 2, 3.

but that was certainly not his emphasis. Roman politics, at least in Arendt's depiction, proceed along similar lines – the Romans glorified those who expanded the empire's reach by conquest or treaties (or treaties in the wake of conquests). For Machiavelli, an interest in glory thus understood was not only to be expected from a political leader; it was the sign of a good leader. It indicated that the appropriate degree of ambition was present. Cicero, in contrast, rejected the idea that glory must be reserved for military affairs. The patient negotiation of political compromises and the creation of successful institutions were no less (and in fact more) important and more deserving of glory than most military triumphs. He worried that military leaders could become focused on glory instead of objectively considering the possible ramifications of a given military project; such misguided priorities would create unnecessary risks for the state. Furthermore, Cicero argued that civic achievements were more permanent than military ones, that political aptitude required as much courage as military exploits, and that it was the competence of politicians that created the resources required for armies to fight in the first place. Cicero's arguments are hardly airtight: It's questionable whether politicians are better at keeping the national interest in mind than generals (consider, for example, the number of times Israeli generals prevented that country's political leadership from embarking on dangerous political adventures). It's also hard to say, categorically, that military achievements are shorter lived or derivative from policy: Military achievements can lay the grounds for policymaking (for example, by obtaining the resources needed for policymaking or by firing up the public's imagination about certain policies), and it is not quite clear that policymaking, even if it occasionally involves personal risk, requires the physical courage that battle does. But even if Cicero's argument is both overstated and self-serving, his fundamental point is worth lingering on: It is easier to glorify a military triumph than a political compromise that sets up retirement funding or expands political enfranchisement. And yet, this facility, the comparative ease of celebrating martial achievements, does not mean that it is also right to prefer warlike pursuits.

Some of the very few contemporary thinkers who think about glory use the term permissively – somewhat similarly to how Cicero did. Timothy Chappell, for example, tells us that the term can attach to any outstanding performance within a worthwhile social practice. His definition of glory as "what happens when a spectacularly excellent performance within a worthwhile form of activity meets the admiration that it merits"[11] is democratic.

[11] Chappell, Glory as Ethical, p. 108.

It applies equally to excellence in cricket and in combat. Stated differently, for Machiavelli, glory is largely coequal with or at least closely related to military glory as the term only has a narrow range of application (military victories and high politics). For Cicero, it should apply to policy attainments as well, and for someone like Chappell, military glory is just one example of glory, one that does not necessarily stand apart or above others. A Bismarck can be glorious for uniting Germany, a fighter in the Battle of the Bulge can be glorious for his valor under fire, and a Tom Brady can be glorious for pulling off an unlikely Super Bowl win – and there's no grounds for comparing these feats because glory resides exclusively within a given activity.

How, then, are we to think about the appropriate scope of glory? To some extent, the answer depends on the psychological function of the concept, on the work we think it should be doing. Here, the Hobbesian take on the desire for glory and transcendence as a psychological risk factor that requires the institution of the state may provide some guidance. If Peg Birmingham's reading of Hobbes is plausible, and the function of glory is to provide ambitious people who have a (dangerously) strong need for recognition an organized outlet for satisfying that need,[12] then it makes sense to define glory narrowly. According to her, the transition from the state of nature to politics is a move from fear of degrading death at the hands of others to the possibility of glorious sacrifice on behalf of the state. According to this interpretation, the Leviathan both makes us safer than we were in the state of nature *and* gives us something potentially glorious to die for. The state answers two human needs, then: security and glory. On this reading, we don't just enter politics because we are afraid to die. We enter politics, among other reasons, because it is the most efficient way to satisfy the itch we have for recognition. Sacrificial violence for the state feeds a natural hunger – for glory and transcendence of one's existence as an individual – a hunger that is far more destructive in the state of nature than it is in politics. The narrower the scope in which glory operates, the greater the degree of recognition it involves, and the greater its power to regulate violent passions. For political glory to sate the need for recognition, there can't be too much of it going around – otherwise, its currency would gradually become diluted. This interpretation prefers the Machiavellian conception of glory to the Ciceronian one; it recommends keeping the term applicable to a limited and especially exalted range of activities so that it can most

[12] Birmingham, Arendt and Hobbes.

efficiently perform its function as a social safety valve.[13] An inflated, permissive idea of glory would leave the superambitious unsatisfied, and the superambitious, Hobbes teaches, are a dangerous bunch.

There is, of course, a simpler way to ask the question: Glory is a form of super recognition. It's meant to honor especially important political achievements. These achievements can be martial. The Allied victory in World War II is a paradigm case. But do they have to be? Why can't civic triumphs qualify? After all, the cementing of the National Health Service is arguably a far greater achievement for the English than their victory over Argentina in the Falklands War. Why not just say that glory should be reserved for recognizing our greatest moments – civic or military? I take this to be Cicero's argument, or at least a generous rendering of it. So, is there a special reason to limit glory to military contexts? Birmingham's interpretation offers one possible answer. But it's hard to test that answer. Would glory really lose some of its powers of psychological regulation if applied to great civic achievements? If exclusivity is what matters, perhaps all we need to worry about is that the term be used sparingly – in whatever context – military or civic. Another reason to limit the term's scope has to do with military recruitment: Armies are (at least sometimes) necessary for the survival of states; the military profession involves danger, it is usually not lavishly remunerated, and, therefore, it should have a greater cut of the glory pie – otherwise we lose an effective way of drawing people of talent and ambition to the armed forces. Again – it's hard to evaluate this argument: Would paying our soldiers a lot solve the problem more effectively? Would there really be a chilling effect on recruitment from conservatively glorifying non-military activities?

Be that as it may, as a matter of fact, glory is still primarily used in its Machiavellian sense – to mark great military achievements and acts of statesmanship. Pericles' soldiers were glorified for defending Athens, England's pilots for winning the Battle of Britain, and Churchill for standing up to the Nazis alone in the first year of World War II. Glory also occasionally comes into play when a nonmilitary project has military implications: Neil Armstrong was glorified for walking on the moon

[13] There is an analogy between the structure of this argument to worries about the dilution of political rights the further they are extended. See, for example, Jacob Mchangama and Guglielmo Verdirame, "The Danger of Human Rights Proliferation: When Defending Liberty, Less Is More," *Foreign Affairs* (July 2013). Related concerns have been raised about humanistic or democratized interpretations of major religious traditions. Thus, does the inclusivity of Unitarian or Reformist understandings of Christianity and Judaism expand access at the risk of diluting the psychological force that stricter compliance to religious traditions exerts on their adherents?

(because landing a man on the moon was a finger in the eye of the Soviets), and Jesse Owens was glorified for beating the German runners in the 1936 Olympics (because his triumph upset the Nazis). But the further away one moves from military or strategic affairs, glory begins to fade from view. The strengthening of social security and the expansion of the right to vote in the United States, Chancellor Angela Merkel's shepherding of European economies out of the 2008 financial crisis, and the Blair administration passing a minimum wage in Britain – all were important political achievements that improved peoples' lives and generated goodwill and even some gratitude from constituents, but they did not bring glory in their wake.

2.4 Glory, Luck, and Time

We have noted that glory is a form of super recognition. As such, its bestowal is outside the control of the glorified subject in several important ways. First, what comes to be glorified is hard to predict. A soldier may be promised glory for fighting in the service of a great cause but the social circumstances may change by the end of the war, generating a consensus that the cause was not so great after all. Fighting in World War II resulted in a glorification of the fighters. Fighting in Vietnam, Iraq, and Afghanistan did not. But often, these social judgments only become evident with time. The soldier receiving a "promise" of glory cannot necessarily know if the cause they fight for will be cherished after the war.

If the social weight attached to the cause one fights for may change in time, so may the broader moral hierarchies that a society holds dear. Even if a soldier or military leader is glorified for their contribution to an important cause, and even if the cause stays stably important, other values that emerge in time can compete with the weight of that cause. Such normative changes can offset the recognition someone has received, and the subject's legacy may change as a result. Thus, as we noted earlier, Churchill was glorified for his pugnacious leadership during World War II. But, in time, as attitudes turned against British imperialism, Churchill's legacy as a colonialist marred, at least for some, his glory as a World War II leader.[14] Similar dynamics have shaped the legacies of historical figures from George Washington to Thomas Jefferson to Bill Clinton – recognition and super recognition – honor and glory – can grow and shrink in time as values shift. Washington and Jefferson's legacies have dimmed as

[14] An example of this critical view can be found in Geoffrey Wheatcroft, *Churchill's Shadow: The Life and Afterlife of Winston Churchill* (New York: Norton, 2021), a revisionist biography of Churchill.

condemnation of their slave ownership went mainstream. Clinton's legacy declined, precipitously, as our views about gender and power have shifted. The heroes of the American Revolution became tainted. The ambitious peacemaker who presided over the Good Friday Agreement and the Dayton Accords, and nearly ended the Israeli-Palestinian conflict, came also to be seen (sometimes primarily) as a creep who did not hesitate to use his position to extract sexual favors.

Finally, and relatedly, our very characters or personality traits can either fit well with the times or be out of joint with them. Machiavelli, as discussed in Chapter 1, noted that Pope Julius was full of impetuous energy and that this quality served him well given the geopolitical circumstances that held during his reign. Then, luckily for his legacy, he died, and his reputation stayed intact, unspoiled by the changed circumstances that would have rendered his personality a liability. One factor shaping how much glory we get, Machiavelli seems to be saying, is whether we live long enough for our tragic flaws to undo us. Julius didn't and that was a big part of why he remained glorious.

2.5 Glory as a Motivator vs. Glory as a Bestowal

A closely related question concerns the difference between glory as a motivational factor and as an ex post facto social bestowal. The prospect of glory often draws young people into battle – as the vision of merging into a larger cause, transcending the smallness and failures of everyday life, and having one's name survive death is enticing. A lust for glory, as both Hobbes and Machiavelli realized, can be dangerous when left unchecked but is also socially useful, indeed indispensable, when harnessed to the goals of the state. And for Clausewitz, it was hard to imagine military motivation – especially among the officer class – without it. As he writes in *On War*:

> Of all the passions that inspire man in battle, none, we have to admit, is so powerful and so constant as the longing for honour and renown. The German language unjustly tarnishes this by associating it with two ignoble meanings in the terms 'greed for honor' (Ehrgeiz) and 'hankering after glory' (Ruhmsucht). The abuse of these noble ambitions has certainly inflicted the most disgusting outrages on the human race; nevertheless their origins entitle them to be ranked among the most elevated in human nature. In war they act as the essential breath of life that animates the inert mass. Other emotions may be more common and more venerated—patriotism, idealism, vengeance, enthusiasm of every kind—but they are

> no substitute for a thirst for fame and honour. They may, indeed, rouse the mass to action and inspire it, but they cannot give the commander the ambition to strive higher than the rest, as he must if he is to distinguish himself.[15]

But whether or not one is actually glorified, whether or not glory is bestowed, is, as we just pointed out, a somewhat contingent social decision. It depends on determinations made by relevant officials about the level of skill displayed in combat, or it depends on whether a society changes its mind about the importance of the cause one went to war for. Thus, a young fighter might be drawn to go to Troy or Agincourt hoping that their performance in future combat will earn them glory. Whether or not it does will be decided by others (formally through medal commissions or informally through word-of-mouth). A young soldier might volunteer to serve in Vietnam thinking that he's going to help face down the rising tide of communism. And yet, if the social consensus that emerges about the cause changes, as it did about the Vietnam War, these hopes for glory will be dashed.

Ernst Junger describes how he and his young German comrades were enticed into the Great War by a romantic vision of fighting: "Surely the war had to supply us with what we wanted; the great, the overwhelming, the hallowed experience. We thought of it as manly, as action, a merry dueling party on flowered, blood bedewed meadows. 'No finer death in all the world than . . .' anything to participate, not to have to stay at home!"[16] As we now know, not much survived of these initial fevered visions. While Junger remained fascinated and even transfixed with the experience of war for the rest of his life, he and his friends did not receive a hero's welcome upon returning – the war had been lost and was, anyways, too long and too exhausting for anyone to celebrate their return.

The social role of glory changes, then, in the life cycle of conflict. Before war begins, the promise of glory will draw people to the fight. From the run-up to World War I to the buildup preceding the war in Afghanistan in 2001, war fever has always been fueled by recruits' and volunteers' visions of glory. Once war starts, combatants often stop caring about glory and start caring about surviving and about each other. At this stage, a preoccupation with glory is seen as vain, irrelevant, and dangerous to the survival of the military team and the success of its mission (the "glory seeker" is a pejorative term among soldiers). Then, after war ends, two further shifts

[15] Carl Clausewitz, *On War*, trans. Howard and Parent (Princeton, NJ: Princeton University Press, 1984): p. 105.

[16] Ernst Junger, *Storm of Steel*, trans. Hoffman (New York: Penguin, 2004): p. 5.

take place: First, at least some of the soldiers who cared nothing for it during combat, begin to demand appropriate recognition for themselves and their fallen comrades. Second, it becomes clear that determinations about whether this recognition will be extended turn on bureaucratic decisions (did X's conduct merit a medal?) and shifting societal views about the meaning of a given conflict. (Was the war in Vietnam really helpful in containing the Soviets? Did we really have any business trying to democratize Iraq?)

Of course, these two areas in which glory operates are not completely distinct: The bestowal of glory can serve as a recruitment strategy for the next generation of fighters. Prospective soldiers can be inspired and moved to fight by seeing how their predecessors are honored. This seems to have been part of the idea behind Pericles' funeral oration – the bestowal of glory on the dead and the promises of immortality and of whitewashing the misdeeds of the fallen were surely calibrated to appeal to the future fighters in the audience. And yet, analytically, the line is clear enough: To be moved by the idea of glory and to actually be glorified are two different states. The first is primarily psychological, the second has to do with the sociopolitical question of what a society decides to value.

2.6 Who Actually gets Glorified?

We might zero in even more closely on the process of glory's bestowal and ask who actually tends to win glory. An unsentimental look suggests that glory is rarely awarded to those who are promised it in the first place. Recall that Periclean glory offers immortality and a degree of transcendence to those who die in order to defend a worthy cause. The whole world is the monument to famous men, Pericles says; the faults and mediocrities of those who die are erased by their sacrifice – wiped clean by the majesty of the idea they upheld. But those who die for such causes are typically not recognized in that way. The world remembers Pericles but none of the infantrymen or sailors who fought under him. The world remembers Lincoln and still marvels at the Gettysburg Address, but the names, lives, and deeds of those who died at Gettysburg are largely lost to us. "The world will little note, nor long remember what we say here, but it can never forget what they did here,"[17] Lincoln famously predicted at Gettysburg, and got it exactly backwards.

[17] Lincoln, https://presidentlincoln.illinois.gov/visit/whats-inside/exhibits/online-exhibits/gettysburg-address-everett-copy/.

"The world" celebrates Churchill, admires his doggedness, his well-timed, beautifully paced speeches, but how many of us can name any of those "few" airmen who fell in the Battle of Britain to whom so many supposedly owe so much?[18] Churchill's earlier crimes of empire may have been whitewashed, at least for a time, and his incompetence at Gallipoli marginalized by the glory of victory over fascism. But how much of that glory flowed downstream? There may well be causes important enough to die for, and those who die defending them may well deserve to be glorified. But they rarely are glorified, and their broken families are more often than not the only ones who remember them. Consider Pink Floyd's "When the Tigers Broke Free" where Roger Waters describes his father's death in the Battle of Anzio during World War II:

And kind old King George
Sent mother a note
When he heard that father was gone
It was, I recall
In the form of a scroll
With gold leaf adorned
And I found it one day
In a drawer of old photographs, hidden away
And my eyes still grow damp to remember
His Majesty signed
With his own rubber stamp
It was dark all around
There was frost in the ground
When the tigers broke free
And no one survived
From the Royal Fusiliers Company Z
They were all left behind
Most of them dead
The rest of them dying
And that's how the High Command
Took my daddy from me[19]

And so, Roger concludes, "[T]he Anzio bridgehead was held for the price of a few hundred ordinary lives"; the promise of glory turned into a

[18] The speech was given on August 20, 1940 and is available here: https://winstonchurchill.org/resources/speeches/1940-the-finest-hour/the-few/.

[19] Roger Waters, "When the Tigers Broke Free" (1983). Lyrics accessed here: https://genius.com/Pink-floyd-when-the-tigers-broke-free-lyrics. Though this 1979 song was originally written for Pink Floyd's album *The Wall*, it was ultimately included in (later releases of) another album – *The Final Cut*.

form letter from King George and a lifetime of hurt for those left behind. A few, like Waters, turn it into art, but most families are just crushed under the weight of it. It's possible that things cannot be otherwise. We may not have the cognitive bandwidth to glorify all who deserve it. It may be easier to simply focus on a Churchill or a Montgomery or a Grant as stand-ins for the great cause, or it may be that, by definition, since glory is a form of super recognition it cannot be widely spread if it is to maintain its prestige and motivational force. But if the latter two observations are correct, there's a moral problem with promises of glory. It's not that glory is an "old lie" as Wilfred Owen claimed in his famous poem.[20] It's not that glory is incoherent in the age of mechanized war (we'll have more to say about this later). It's that the promise of glory cannot be kept on anything like the terms on which it's typically made. It's that glory is almost always oversold and underdelivered.

2.7 Refracted Glory

Abstracting for the moment from the question of whether it is the people who are most deserving that actually get glorified, it is worthwhile considering if this form of recognition travels. Can those who were not related at all to a glorious deed be recognized for it? And can a nation bask in the glory of those who risked their lives for it?

Thus, for example, could an American retiree in the summer of 1944 be glorious because of the deeds of those who landed on the beaches of Normandy? And does the glory of those who landed there somehow attach to the nation itself? Would the glory of the Apollo astronauts attach to an American doctor or school teacher watching the moon landing on TV? And do the feats of Armstrong, Aldrin, and their colleagues somehow make the nation itself glorious? It seems wrong to suggest that the doctor or the teacher is glorious for something they were not personally involved in. And it is coherent, but a bit awkward, to suggest that these deeds made the nation glorious. Perhaps the more meaningful description is that important activities that earned people glory can contribute to a sense of national honor – a feeling that both attaches to individuals and enhances national reputation. Americans would appropriately feel pride for D-Day – both during the summer of 1944 and later on. And the reputation of the United States would be enhanced by that military success. The achievements that win individual people glory generate national pride and

[20] Owen, *Collected Poems.*

that pride can make individuals proud to be part of the nation. But glory itself is, strictly speaking, a form of recognition that properly attaches to individuals. A cause can be great, a nation can be worth defending, but glory is due to those who served them. In other words, the results of glorious deeds will often accrue to national honor and national honor can be felt by ordinary citizens – even over an expanse of time and space. But glory itself is limited to those directly involved. That conclusion seems particularly apt if we accept the Hobbesian view of glory as a disposition meant to regulate ambition and harness it to the service of the state. As we stated earlier, to achieve this function, glory must be applied narrowly and conservatively.

2.8 Glory and Duty

The practice of glorifying warriors (or at least the promise to do so) sits uncomfortably with the assumption that soldiers have an obligation or duty to fight for their country.

If a soldier has a duty to fight for an important cause, should they also be glorified for doing so? Does it make sense to glorify someone for what they *ought* to do?[21] The idea of duty usually implies that one can be criticized (and sanctioned) for failing to comply while one is not praised (and certainly not glorified) for adherence.[22] A parent who provides their child with clothes and food is fulfilling their duty. They should be criticized if they fail to perform these basic tasks, but it would be absurd to praise them when they carry them out. And yet, we do regularly glorify (or at least promise to glorify) those who are doing their duties: Churchill is glorified for being a good prime minister during World War II, but he was *supposed* to be a good prime minister during World War II. Those who fought and died on the beaches of Normandy are glorified though they were under a military obligation to fight. What explains this tension?

One possibility is that we glorify people for performing their duty exceptionally well – better than can be expected. On this explanation, glorifying Churchill as a wartime leader, or promising to glorify the RAF fighter pilots, was based on the fact that these people carried out an obligation with unusual competence. Another possibility (not necessarily mutually exclusive with the first one) is that the bestowal of glory is

21 I am grateful to my colleague Chris Zurn for this observation.

22 For an interesting discussion of the distinction between charity and duty, see Peter Singer, "Famine Affluence and Morality," *Philosophy and Public Affairs* 1(3) (1972): 229–43.

psychologically rather than morally motivated: One does not morally deserve glory for fulfilling a martial duty but one should still be granted it because duty alone is not a strong enough motivational factor to draw soldiers into war. Duty is one thing, conjuring up the motivation to act on one's duty is another. To promise glory or to glorify is to psychologically attach people to their duties. The practice of glorifying, on this understanding, is simply a reflection of the fact that duties are not self-executing. As both Burke and Machiavelli argue, in very different contexts, duties require motivational support. For Machiavelli, important legal institutions must be buttressed by religion if they are to be taken seriously: "Although the good things known to a prudent man are many, these things in themselves lack the self-evident qualities that can persuade others."[23] Burke echoes a similar insight in the *Reflections*. We don't follow laws because of the truth value and plausibility of what they say but because of a historical and psychological structure that makes them seem worthy of adherence: "[P]ublic feelings, combined with manners, are required sometimes as supplements, sometimes as correctives, always as aids to law. The precept that a wise man who was also a great critic gave for the construction of poems is equally true of states: 'It is not enough that they be beautiful; they must also be persuasive'. Every nation should have a system of manners that a well-informed mind would be disposed to enjoy. To make us love our country, our country ought to be lovely."[24] Glory, on this Burkeian understanding, makes duties more "lovely."

2.9 Internal and External Explanations of Glory

The two modes of explanation just described – seeing glory as a reward for performing a duty particularly well and viewing it as a psychological inducement needed to fulfill a duty – proceed from two different perspectives. The first is internal and the second external to a given social practice. And that same difference plays out across the historical accounts offered in Chapter 1. Let us explain in some detail.

The Periclean account proceeds from what we may call an internal point of view – the funeral oration is the story of a society and a leader glorifying what they see as very important. For Pericles, the cause his soldiers died for was so important that the fighters deserved glory and all of its attendant

[23] Machiavelli, *Discourses on Livy*, I.11, p. 52.

[24] Edmund Burke, *Reflections on The Revolution in France* (Indianapolis, IN: Hackett Publishing, 1987): p. 68.

benefits. This is glory described from the inside – from the point of view of participants in a social practice. For those participants, their love of Athens is *a reason* to fight (rather than a prediction that citizens probably will fight) and the glory promised in return for fighting makes sense as part of the expectations inherent to that practice. Hobbes' account of glory, in contrast, proceeds from the external point of view: It is, to be anachronistic, a social-scientific explanation – human beings are so preoccupied with recognition that something like the official regulation and bestowal of glory by the state is necessary for coordinating their competitive tendencies. For this mechanism to work, of course, the participants must relate to the promise of glory as internally significant. But that does not make it anything other than it really is, viewed from the outside: A social mechanism necessary for creatures like us to live together.[25]

Machiavelli and Arendt's accounts are mixed: Each provides us with an external explanation of the social function of glory (for Machiavelli a necessary driver of any effective leader, for Arendt a value characterizing political systems preoccupied with achieving permanence through great works rather than managing the basic needs of large groups). But both thinkers also have an internal and normative attachment to the term. Machiavelli's project argues for the goodness of *gloria mundi* – of political leaders and states pursuing earthly glory. And it's hard to miss the nostalgic tone in Arendt's argument – as far as she's concerned, something important was lost in the transition from Greek politics and its preoccupation with freedom from necessity and lasting works to modern mass politics and its preoccupation with mass administration or "society."

2.10 The Death/Glory Nexus

Does glory somehow depend on the death of its subject? The historical accounts we surveyed reveal some important and inconsistent answers. Pericles' initial hesitation to deliver his oration notwithstanding, glorifying the dead is relatively straightforward. As he conceded, departed heroes don't compete with anyone for power and therefore it is easier to lavish praise on them than on the living. As we saw earlier, Machiavelli warns that glorifying the living – a conquering general for example – presents serious difficulties. A surviving victor can be a direct threat to existing leaders – because they attract too much popular support or because their

[25] For a classical discussion of the nature of the two points of view, see Herbert Hart, *The Concept of Law* (New York: Oxford University Press, 1994): chapter 4.

glory increases their own ambition to rule. Either way, their relationship with the ruling class can easily become tense. One thinks here of the contentious relationship between Truman and MacArthur, Begin and Sharon, and countless other historical cases where a general's fame threatens to upstage a leader's political standing.

Furthermore, the reputation of the glorious dead is, at first glance, more secure than that of those who are glorified during their lifetime, because the dead can't ruin it. One thinks here of Rudy Giuliani, who provided real leadership as mayor of New York City in the aftermath of the September 11 attacks only to blow up his legacy in later years. The Hebrew phrase "his old age shames his youth" captures the basic point – to die gloriously soon after the deed one was glorified for is as close as we can get to a guarantee of long-lasting reputation. We've complicated this picture earlier by arguing that the glory of the deceased may shift in time as a result of changed views of the project they died for, or of other activities the dead were engaged in during their lifetime. And it's also worth mentioning that the living don't necessarily have to ruin their reputation. They can, and often do, actively spend time on curating it. The presidential libraries of American presidents are, essentially, institutions devoted to such curation. And both Churchill and Stalin reportedly offered variations on the idea that history would be kind to them because they would help to write it.[26]

Sophocles offers another take on the relationship between glory and death in *Ajax*: dying itself can serve as a tool for reputation management – in the case of Ajax, a way of stopping humiliation at the hands of rivals and perhaps even reinstating some of his past glory.[27] Something like this is true across cultural contexts – dying can either actively repair one's reputation or diminish the motivation of others to attack it. Examples vary from suicide killers trying to clear themselves from charges of disloyalty[28] to officials who commit ritual suicide to atone for a perceived misstep

[26] See, for example, David Reynolds, *In Command of History: Churchill Fighting and Writing the Second World War* (London, UK: Allen Lane, 2004): p. 639, and Richard Langworth, *Churchill in His Own Words* (London, UK: Ebury Press, 2012): p. 64.

[27] Tecmessa says of Ajax's demise: "His Death was as bitter to me as it's sweet to them; but for himself a happiness for he won his great desire, the death he looked for" in Sophocles, *Ajax*, trans. Jameson (Chicago, IL: The University of Chicago Press, 2013): lines 966–68.

[28] For one example, see Ron Dudai and Hillel Cohen, "Triangle of Betrayal: Collaborators and Transitional Justice in the Israeli-Palestinian Conflict," *Journal of Human Rights* 6(1) (2007): 37–58. doi.org/10.1080/14754830601084642. The authors write: "Palestinian society has thus far largely failed to devise reintegration paths, or 'exit strategies,' for those who are associated—this way or the other—with the stigma of collaboration. It seems like one of the few existing paths is that of becoming engaged in violent attacks against Israel or becoming a 'martyr'" (p. 49).

and salvage their honor.[29] The point here is not so much that the dead are easier to glorify or that their legacy is more stable, but that sometimes dying itself offers reputational benefits. It is, however, easy to overstate this argument. As Teucer, Ajax's half brother, reminds us, the dead are especially susceptible to humiliation: "[W]hen a man lies dead and helpless, all the world delights to mock and injure him."[30] King Menelaus, for one, was bent on humiliating Ajax by refusing him a burial: "[H]e shall be tossed forth somewhere on the pale sand, to feed the seagulls."[31]

2.11 Normative Considerations

We have, so far, focused on conceptual questions pertaining to the nature of glory. But if these considerations help to make the idea of glory more philosophically coherent, we are still left with questions about the normative application of the term.

Recall that we are working with two concepts of glory – the political or Periclean variety that acknowledges the willingness to take risks for important causes, and the amoral Achillean kind of glory that turns primarily on recognizing skill. Let's begin with the normativity of Periclean glory. How are we to think about questions of *ought* here? On the most straightforward level, we might evaluate glory in the same way we reflect on the justice of fighting for the political cause that gave rise to glory in the first place. In other words, if the cause cannot pass *jus ad bellum* muster – if it cannot ground a just war – then fighting for it cannot result in glory (or, glorifying someone for fighting in the name of it would be morally wrong for the same reasons that fighting the war would be). Thus, for example, glorifying a Wehrmacht officer for serving the Nazi cause would be morally wrong. His glory would be ill-assigned and undeserved. On the same token, an Allied officer or a Union commander would be rightly glorified on jus ad bellum grounds for their roles in World War II and the Civil War, respectively.

Moving to the second level of classical just war theory, one should not win glory for a campaign during which they jettison basic *jus in bello* precepts such as the need to observe proportionality or noncombatant immunity. Thus, Sir Arthur Harris, aka "Bomber Harris," Air Chief Marshal for Great Britain during World War II, deserves no glory for his

[29] For some interesting examples and an overview, see Mark Hastings, Lisa Northman and June Tangney, "Shame, Guilt, and Suicide," in *Suicide Science* (Boston, MA: Springer, 2002).

[30] Sophocles, *Ajax*, line 985.

[31] Ibid., line 1060.

insistence on the "area bombing" campaign that intentionally targeted civilians in German cities. According to Harris:

> The aim of the Combined Bomber Offensive . . . should be unambiguously stated [as] the destruction of German cities, the killing of German workers, and the disruption of civilised life throughout Germany . . . the destruction of houses, public utilities, transport and lives, the creation of a refugee problem on an unprecedented scale, and the breakdown of morale both at home and at the battle fronts by fear of extended and intensified bombing, are accepted and intended aims of our bombing policy. They are not by-products of attempts to hit factories.[32]

The way Harris fought the war violated fundamental *jus in bello* principles and, as a result, he shouldn't be glorified for it. There's not much to say about the normativity of Achillean glory. As we have insisted earlier, it's primarily amoral. But the picture is a bit more complicated. Recall that Achillean glory is focused on fighting skill – on how rather than why one fights and that, as a result, this idea of glory is fundamentally amoral. But the "technical" emphasis of Achillean glory, its very emphasis on the manner in which a martial task is carried out, may have some surprising moral benefits. Take military families who pride themselves on a tradition of fighting professionally. Aren't soldiers who are committed to such an ethos more likely to economize on the costs of war? Stated differently, would a soldier concerned with their family's legacy of professionalism be more or less deadly than one who is concerned with the sanctity of a political cause – a "war to end all wars," as World War I was often billed, or a war to defend political freedom as World War II and the post-September 11 wars in Afghanistan and Iraq were described? As I have argued elsewhere, soldiers and military leaders can be blinded by the magnitude or "sanctity" of great political causes.[33] Indeed, that perception of sanctity may justify a great deal of slaughter. Here's how Viscount Bryce, a prominent British diplomat, described the Great War in a gathering he hosted to celebrate the publication of a book about German militarism: "We are fighting for great principles. Principles vital to the future of mankind, principles which the German government has outraged and which must at all costs be vindicated to defeat militarism. This is a conflict for the principles of right which were violated when

[32] Arthur T. Harris, *Dispatch on War Operations: 23rd February 1942 to 8th May 1945* (London, UK: Frank Cass, 1995), p. 7.

[33] Nir Eisikovits, *A Theory of Truces* (London, UK: Palgrave MacMillan, 2015): chapter 1.

innocent non-combatants were slaughtered in Belgium and drowned on the Lusitania. The allies are bound and resolved to protect it till a victory has been won for these principles and for a peace established on a sure foundation of justice and freedom."[34] Lofty aims, no doubt. If they could be achieved, if they were within reach, they could justify a great deal of killing and promise a great deal of glory to those who die for them. But positing these kinds of causes rather than more limited and less lofty ones has, historically, often both prolonged and escalated wars. One can, to paraphrase a famous saying, talk oneself into breaking many eggs in order to prepare a choice omelet. The aims of the professional soldier are more modest. They want to fight well; they want to make sure they don't embarrass themselves. If there's a family legacy, they will often worry about keeping it intact. They will not easily step out of their zone of expertise for fear that such adventurism may undermine these goals. In short, there is – not conceptually but often practically – a conservatism built into the conduct of those interested in Achillean glory that is often lacking in those interested in Periclean glory.[35] The professional warrior is part of a guild and very often also part of a family tradition, both of which make them better at their military work and, at the same time, impose restrictions and norms of behavior. And this very conservatism may end up economizing on the harms of war.

Ironically, Achilles himself does not provide the best example of this disposition, animated as he so often is by rage. In her book *The Code of the Warrior: Exploring Warrior Values Past and Present*, philosopher Shannon French tells us that her students at the US Naval Academy regularly chose the Trojan prince, Hector, as their model for the consummate professional warrior, because he was "a man who fights with tremendous ferocity on the battlefield but who is not driven by rage or bloodlust." Hector, guided as he was by a sense of obligation to his city and family (but certainly not to the political cause he was fighting for – he cared nothing for defending his brother, Paris), stood out to these young professional soldiers as a role model.[36]

Professional warriors typically operate on the basis of a code, and that code serves to limit what they can do, to distinguish them from cold-blooded killers, and even to protect them from the psychological ravages of war. Such a code, French writes, "defines not only how a warrior should interact with their own ... comrades but also how they should treat other members of their society, their enemies, and the people they conquer. The

[34] Ibid., p. 5. [35] I am again indebted to Chris Zurn for interesting conversations on this point.
[36] Shannon French, *The Code of the Warrior: Exploring Warrior Values Past and Present* (Lanham, MD: Rowman and Littlefield, 2003): p. 31.

code restrains the warrior. It sets boundaries on his or her behavior. It distinguishes honorable acts from shameful acts. Under the codes of chivalry, a medieval knight has to offer mercy to any knight who yields to him in battle. In feudal Japan, samurai are not permitted to approach their opponents in stealth but rather are required to declare themselves openly before engaging in combat. Muslim warriors who are engaged in offensive jihad cannot employ fire as a weapon."[37]

The same picture of professional soldiering being limited by rationality, judgment, and the requirements of the military profession emerges from philosopher Nancy Sherman's important account of what stoicism can teach us about fighting in war. Of particular interest in this context is her discussion of the benefits of military decorum (from the wearing of uniform to posture and norms of address) for operational efficiency and control. The main point, supported by sources as diverse as Erving Goffman and Kant, is that rules and expectations of military decorum, so central to the lives of soldiers, and reinforced by a tradition of professional military service, serve to instill norms that guide and limit how soldiers act. Putting on the clothes, expressions, manner, and body postures of soldiers helps them to adhere to a military ethic. That this counts as "acting" does not detract from the importance of these rituals – after all, most instances of role morality are supported by acting. Sherman writes: "Military rules and regulations, and what can fall under the category of 'conduct unbecoming,' pertain to actions, but they also cover a more general sense of comportment, manners, and emotional bearing. Whether or not it is a throwback to Roman and medieval times, or the replacement of Homeric anger and lust with a chivalrous warrior's code of downcast eyes and stolid demeanor, an emphasis on the face and look of a warrior remains with us."[38]

This (very Aristotelian) argument is straightforward: Military decorum and the soldiers' code are helpful in building character and habits. As such, regardless of their intrinsic value, they have a moderating and ultimately salutary effect on soldiers' behavior.[39]

2.12 Summary

Let us take stock: Glory is a particularly elevated form of honor, a kind of "super recognition." It is more exclusive and longer lasting than honor,

[37] Ibid., pp. 4–5.
[38] Nancy Sherman, *Stoic Warriors: the Ancient Philosophy Behind the Military Mind* (New York: Oxford University Press): p. 43.
[39] See in particular, Ibid., chapter 3.

and it is typically connected with promises of immortality and an "upgrade" of one's reputation (as it diminishes the importance of past misdeeds). We have distinguished between political (or Periclean) and personal (or Achillean) glory. The former is awarded for a willingness to sacrifice oneself for an important cause and does not turn on the level of skill displayed. The latter is awarded for excelling others in activities that (typically) involve danger. This kind of glory is largely indifferent to the political cause involved and does turn on ability. Personal glory is competitive by definition, political glory is not. We have also discussed the scope of the term and have suggested that determining the proper objects of glory (military, political, cultural, or even everyday pursuits) turns on the social role the concept is supposed to play. We noted that glory (much like reputation more broadly) is subject to changes in moral norms over time. Stated a bit differently, whether, or how long, or how much one is glorified is subject to what philosophers call "moral luck" – particularly luck in how the political cause one fought for fares in time and how other aspects of one's behavior come to be seen by later generations. To the extent that glory is subject to luck in this way, the Periclean promise tying glory to everlasting fame cannot be fully realized. We also noted that the status and role of glory change during different stages of a conflict or political project. Early on (typically before a war starts) glory serves as an aspiration, a hope motivating people to fight for the cause. During the conflict, the preoccupation with glory usually fades among those who actually do the fighting, and after the conflict, the question of bestowing glory becomes subject to bureaucratic and social decisions. The glory bestowed is not necessarily congruent with the kind of glory that individual soldiers envision when they sign up. Furthermore, we noted that often those who actually do the fighting are not the ones who get glorified. Contrary to the iconic rhetoric contained in speeches like Pericles' funeral oration, Lincoln's Gettysburg Address, or Churchill's World War II speeches, it tends to be leaders (who were famous to begin with) who get the lion's share of the recognition. We noted the tension between positing that someone has a duty to fight and the practice of glorifying them for fulfilling that duty and we also argued that glory is subject to both internal and external explanations. The former kind of explanations proceed in terms that are meaningful to participants in a given social practice, and the latter are focused on the function of the term as a social driver of behavior. We concluded by tracing the relationship between glory and death, and examining the normativity of both Periclean and Achillean glory.

CHAPTER 3

Our Ambivalence about Glory

Glory may have become passé for political philosophers, but we still hear quite a bit about it in public discourse. When we do, what we hear is shot through with ambivalence. Often, glory is taken as seriously as it was considered by Pericles. President Zelensky, for example, speaks of "slava Ukraini" or "glory to Ukraine,"[1] much in the same way the Athenian leader spoke about defending his city in the fifth century BCE: Risking all for a great cause will win those who make the sacrifice gratitude and great recognition. On the other side of the moral spectrum, ISIS recruiters also speak of glory accruing to those who would help bring back the historical Caliphate and reverse centuries of humiliation of Muslims by "crusaders" or apostate Muslims[2]. In contrast, war-weary skeptics regularly argue that the term is a sham, a thinly veiled strategy for making the young fight the wars of the old: "Your 'fight them on the beaches' speeches make me despair 'Cause if there's one thing we can guarantee is you will not he there," sings The Beautiful South[3]. And Wilfred Owen, the best known among the World War I poets, famously calls the idea of a glorious death for the "patria" an "old lie."[4]

How are we to understand these tensions? Why has glory become philosophically passé even though it is still ubiquitous in politics? And, more importantly, why are political views of the term so discordant? In this chapter, I'll try to answer these questions by examining some of the key sources of skepticism about martial glory against the theory developed in the first two chapters.

[1] See for example here: www.youtube.com/watch?v=j9bpS9MXeDA.

[2] For a detailed discussion, see Chapter 7.

[3] The lyrics are from The Beautiful South, "Have You Ever Been Away," *Welcome to the Beautiful South* (1989).

[4] Owen, *Collected Poems*.

3.1 The Arguments against Glory

3.1.1 Glory Kills, Commerce Preserves

In his influential account of the rise of capitalism, the historian Albert Hirschman charts how commerce and the quest for wealth came to replace military pursuits as markers of greatness in the last few centuries. Interest in martial glory receded as it became clear that the aristocratic culture tying honor with war was inherently dangerous. Commerce, the early liberal orthodoxy held, had the potential to curb war's excesses. Replacing our fixation with honor with an ethos of moneymaking was the key to peace and stability.

Crucial to making the transition, according to Hirschman, was deploying the *principle of countervailing passion* – pitting one set of psychological needs against another in order to change behavior. As it took shape, this strategy settled on pitting peoples' interests (understood to cover material welfare[5]) against their preoccupation with honor, with the assumption that it is safer to pursue the former. Rather than improving their situation by attaining more glory, the "augmentation of fortune [became] the means by which the greater part of men propose and wish to better their condition," as Adam Smith put it.[6] Thus, the pathway to social improvement was democratized, and the passion for money would replace and limit the passion for honor – "one set of passions hitherto known variously as greed avarice or love of lucre, could be usefully employed to oppose and bridle such other passions as ambition, lust for power, or sexual lust."[7] Avarice, which until the seventeenth century had been considered a deadly sin, was "rebranded" as the pursuit of one's self-interest – a preoccupation which, while less than chivalric, is conducive to peace. In fact, the new thinking went, such selfish pursuits have the potential to promote, taken in aggregate, the general welfare. As Mandeville put it perfectly in his famous *The Fable of the Bees*, "[E]very Part was full of Vice, Yet the whole Mass a Paradise . . . Such were the Blessings of that State; Their Crimes conspired to make 'em Great."[8]

[5] Albert Hirschman, *The Passions and the Interests: Political Arguments for Capitalism Before its Triumph* (Princeton, NJ: Princeton University Press, 2013): p. 38.

[6] Ibid., p. 40. [7] Ibid., p. 41.

[8] Bernard Mandeville, *The Fable of the Bees or Private Vices, Public Benefits* (London, UK: T. Ostell, 1806 – via the Gutenberg Project): line 155. www.gutenberg.org/cache/epub/57260/pg57260-images.html.

The pursuit of self-interest emerged as a third category between passions and reason – a preoccupation free from the explosiveness of the former, but also a more effective and potent motivator for action than the latter.[9] Peoples' material interests, on this view, were not only a safer way to get them to act than to send them on a quest for honor, they were also more predictable and more constant. It was easier to figure out what someone else's interests were and to anticipate how they would act to satisfy those interests than it was to predict what would make someone feel honored or slighted. Elevating the pursuit of economic interest thus became key to domestic and international tranquility: individuals and nations that focused on augmenting their wealth were more moderate and predictable than those animated by honor and glory. "Avarice is a uniform and tractable vice," Samuel Johnson's Rasselas insists, adding, "other intellectual distempers are different in different constitutions of mind; that which soothes the pride of one will offend the pride of another but to the favor of the covetous there is a ready way: bring money and nothing is denied."[10]

The turn to money as the marker of recognition was, in some fundamental way, born of trauma – the pursuit of personal and national honor yielded so much death and destruction that another focus was required. Sure, an interest in money could be contemptible, but it was also safer. As Hirschman puts it, "[I]n an age in which men were searching for ways of limiting the damage and horrors they are wont to inflict on each other, commercial and economic activities were ... looked at more kindly not because of any rise in the esteem in which they were held; on the contrary any preference for them expressed a desire for a vacation from (disastrous greatness)."[11] Capitalism, on this picture, arose as a pacification strategy moving what was once contemptible (moneymaking) into the mainstream because the old ways of pursuing respect became too dangerous to sustain.

The culmination of this intellectual trend is, of course, Montesquieu's classical argument about the link between commerce and peace – his claim that extensive mutual trade and the economic interdependence that it yields make war unthinkable: "[T]he natural effect of commerce is to lead to peace. Two nations that trade together become mutually dependent: if one has an interest in buying, the other has one in selling; and all unions are based on mutual needs."[12] A variation on this theme suggests that allowing a complicated modern economy to develop precludes heavy-handed and ill-informed domestic actions by leaders bent on personal or

[9] Hirschman, *Passions and Interests*, pp. 43–44. [10] Ibid., p. 55. [11] Ibid., p. 59.
[12] Ibid., p. 80.

national glory – simply because the functioning of a modern economy is subject to too many interdependencies and too many unpredictable variables to accommodate this kind of intervention. A ruler who did not want to blow up their own economic system (and damage their reputation in the process) would have to wield power cautiously. The very act of setting up a modern economy is tantamount to accepting restrictions on the exercise of power: "The power of a modern prince, let it be by the constitution of his kingdom, ever so absolute, immediately becomes limited so soon as he establishes the plan of oeconomy which we are endeavouring to explain," wrote Sir James Steuart. "If his authority formerly resembled the solidity and force of the wedge (which may indifferently be made use of, for splitting of timber, stones and other hard bodies, and which may be thrown aside and taken up again at pleasure), it will at length come to resemble the delicacy of the watch, which is good for no other purpose than to mark the progression of time, and which is immediately destroyed, if put to any other use, or touched with any but the gentlest hand. As modern oeconomy, therefore, is the most effectual bridle ever-...invented against the folly of despotism so the wisdom of so great a power never shines with greater lustre, than when we see it exerted in planning and establishing this oeconomy, as a bridle against the wanton exercise of itself in succeeding generations."[13]

But if the rise of a complicated commercial society weakens the martial spirit and curtails despotism, it also exacts a cost in civic virtue. Adam Smith agreed with Montesquieu about the moderating effect of commerce, but he was concerned about its psychological costs. The citizen in commercial society, he argued, became diminished in important ways: "[O]f the great and extensive interests of his country, he is altogether incapable of judging; and unless very particular pains have been taken to render him otherwise, he is equally incapable of defending his country in war. The uniformity of his stationary life naturally corrupts the courage of his mind, and makes him regard with abhorrence the irregular, uncertain, and adventurous life of a soldier."[14] In other words, the division of labor and the focus on luxuries dampen thumos: "[C]ommerce ... sinks the courage of mankind, and tends to extinguish martial spirit. In all commercial countries the division of labour is infinite, and every one's thoughts are employed about one particular thing ... In the same manner war comes to be a trade also. A man has then time to study only one branch of business, and it would be a great disadvantage to oblige everyone to learn the

[13] Ibid., p. 84. [14] Ibid., p. 106.

military art and to keep himself in the practice of it. The defense of the country is therefore committed to a certain set of men who have nothing else ado, and among the bulk of the people military courage diminishes."[15]

In a similar fashion, thinkers like Ferguson, Tocqueville, and later, Marx and Durkheim note that social bonds in a commercial republic fray and solidarity is weakened because each person in these states "deals with his fellow creatures as he does with his cattle and soil – for the sake of the profit they bring."[16] A functional economy that secures tranquility is "not just a bar to princely caprice"[17] it is also an inducement to stability – sometimes too much stability – a conservatism that eschews rocking the boat. The demand commercial classes make for law, order, and constancy can introduce paralysis and even grounds for a new kind of authoritarianism into politics: "[I]f it is true that the economy must be deferred to, then there is a case not only for constraining the imprudent actions of the prince but for repressing those of the people, for limiting participation, in short for crushing anything that could be interpreted by some economist-king as a threat to the proper functioning of the 'delicate watch'."[18]

The accelerated rise of capitalism after the Congress of Vienna (in 1814–1815) gave rise to a romantic critique that amplified these concerns: The mercantile world and the merchants in it were boring, soulless, and without honor, passion, or nobility. But that was, for the proponents of the *commercial peace* exactly the point: The qualities that the romantics celebrated had, in the past, proved deadly. Let critics of capitalism complain about its deadening effect on personality, let them wail about commercially induced uniformity. Boring and alive is better than glorious and dead. "Capitalism," Hirschman tells us, "was supposed to accomplish exactly what was soon to be denounced as its worst feature."[19]

By turning the vice of greed into a virtue, early modern and later enlightenment thinkers argued that the aristocratic culture of honor and its attendant valorizing of war could be defanged. Some of these thinkers were more ambivalent about this process than others (as the anxieties about the cost of capitalism for civic-mindedness just outlined attest) and some saw wealth as merely another avenue to pursue glory.[20] But all

[15] Ibid., p. 106. [16] Ibid., p. 120. [17] Ibid., p. 122. [18] Ibid., p. 124.
[19] Ibid., p. 132.
[20] In *The Theory of Moral Sentiments*, Smith challenges the idea that an interest in money proceeds from different psychological origins than the need for honor. He argues that it is the need to be noticed rather than the desire to make one's life comfortable that drives people to accumulate wealth. "[T]o be observed, to be attended to, to be taken notice of . . . are all the advantages which we can propose to derive from [wealth]. It is the vanity, not the ease or the pleasure which interests

of them shared the basic belief that a politics that ran on money – though in many ways less attractive – would prove more stable and more peaceful than one that ran on glory.

This classical explanation for glory's long hibernation is grounded, then, in fear. Anxiety about the damage that centuries of preoccupation with recognition and renown had wrought led to the turn from martial glory to commerce. To state the obvious, this classical argument for commerce has not stood the test of time. The last century and a half have provided ample evidence that commercial interests are easily translated into military ones, that economic interdependence is not a guarantee against war, and that the inequality and alienation attendant to unfettered free markets can easily destabilize both domestic and international relations.

3.1.2 *The Old Lie: Glory after World War I*

So much for the classical enlightenment argument against glory. If the idea has become particularly suspect in the last century, that suspicion is due, in large part, to the legacy of World War I. In his account of the cultural ramifications of that war, Paul Fussell describes the heartbreaking story of English infantrymen who, early on in the conflict, kicked soccer balls into No-Man's-Land just before beginning an attack. For Fussell, this attitude, this propensity to view war as a game, had its roots in the relatively peaceful century that preceded the Great War. He quotes from a poem describing this "sporting" spirit:

On through the hail of slaughter,
Where gallant comrades fall
Where blood is poured like water
They drive the trickling ball
The fear of death before them
Is but an empty name
True to the land that bore them
The Surreys play the game.[21]

One notes the passing reference to "slaughter" or the blood flowing "like water," but these are primarily abstractions or, to use the poet's language, features of the "game" that was war in the young men's romantic

us," Smith, as quoted in Hirschman, *Passions and Interests*, p. 108. On this view, far from softening the need for recognition and honor, the pursuit of profit becomes a new vehicle for it.

[21] Paul Fussell, *The Great War and Modern Memory* (New York: Oxford University Press, 2000): pp. 27–28.

preconception. Later, after the carnage in the Somme, Verdun, and Passchendaele, the perception of fighting (and with it the war poems) grew bleaker. There was no sportsmanship, chivalry, or glory to be had when men were being mowed down, blown up, and gassed. War as sport gave way to war as industrial slaughter. For the poet Wilfred Owen, who was killed in France a week before hostilities ended, martial glory could not survive the way the Great War was fought. It was grotesque to speak about glory when people died as his comrades did. The last stanza of his *Dulce et Decorum Est* has become an iconic representation of this sentiment:

> If in some smothering dreams, you too could pace
> Behind the wagon that we flung him in,
> And watch the white eyes writhing in his face,
> His hanging face, like a devil's sick of sin;
> If you could hear, at every jolt, the blood
> Come gargling from the froth-corrupted lungs,
> Obscene as cancer, bitter as the cud
> Of vile, incurable sores on innocent tongues,—
> My friend, you would not tell with such high zest
> To children ardent for some desperate glory,
> The old Lie: *Dulce et decorum est*
> *Pro patria mori.*[22]

Talk of military glory was worse than the excitable, sentimental drivel offered by those who had not seen real war. For Owen, to keep talking about glory after seeing what modern war did to soldiers was to peddle a pernicious lie calculated to convince young men to die in a fashion that could not possibly be heroic. Where there is industrial slaughter, where men do not die as individuals but are killed as cannon fodder, fed into the great machine of total war, the very idea of glory becomes either obscene or incoherent.

Frederic Henry, Ernest Hemingway's protagonist in *A Farewell to Arms*, makes a very similar point about his contempt for abstract descriptions of war: "I was always embarrassed by the words sacred, glorious, and sacrifice and the expression in vain. We had heard them, sometimes standing in the rain almost out of earshot, so that only the shouted words came through, and had read them, on proclamations that were slapped up by billposters over other proclamations, now for a long time, and I had seen nothing sacred, and the things that were glorious and had no glory and the sacrifices were like the stockyards at Chicago if nothing was done with

[22] Owen, *Collected Poems*.

the meat except to bury it. There were many words that you could not stand to hear and finally only the names of places had dignity. Certain numbers were the same way and certain dates and these with the names of the places were all you could say and have them mean anything. Abstract words such as glory, honor, courage, or hallow were obscene beside the concrete names of villages, the numbers of roads, the names of rivers, the numbers of regiments and the dates."[23] The very deployment of terms like glory to describe war, Hemingway seems to suggest, words which are meant to elevate and sanctify the causes which soldiers fight for, are blasphemous and sacrilegious to them. The only terms that carry weight with them, the only phrases which have a kind of secular gravity for them, are those denoting the concrete experiences of fighting.

3.1.3 Asymmetrical War and the Possibility of Glory

It's not just the impersonal, grotesque, industrial nature of dying in modern combat that raises doubts about military glory. The idea has come under pressure from other quarters as well in recent decades. Since World War II, much intergroup conflict has become asymmetrical. The face of war, to use two cinematic paradigms, has, with some important exceptions, shifted from "Saving Private Ryan" to "The Battle of Algiers." When fighting takes place between a powerful organized army on the one hand and weaker guerilla forces who must use civilian cover if they are to stand a fighting chance on the other, one begins to move from the possibility of "fighting well" to General Sherman's old edict about war being "hell," where extreme and widespread suffering by innocents is unavoidable.[24]

In one of the famous scenes in "The Battle of Algiers," a movie about the Algerian War of Independence against the French, the leader of the French troops, Colonel Mathieu, reminds a group of journalists questioning the torture of captured FLN fighters that the entire political spectrum in France wanted to keep Algeria French. "The question is this," the Colonel insists: "Should France stay in Algeria? If your answer is yes,

[23] Ernest Hemingway, *A Farewell to Arms* (New York: Scribner, 2014): chapter 27.

[24] The more precise sentiment is captured in Sherman's September 12, 1864 letter to the aldermen of Atlanta who protested his plan to evacuate and burn down their city: "[Y]ou cannot qualify war in harsher terms than I will. War is cruelty and you cannot refine it; and those who brought war into our country deserve all the curses and maledictions that a people can pour out." https://cwnc.omeka.chass.ncsu.edu/items/show/23.

you should accept all the necessary consequences."[25] Given the nature of the fight (an army facing off against small, highly independent guerilla units), there is no way to fight the FLN "with honor." The only moral decision to be made was whether to fight at all. In a later scene, the FLN leader, Larbi Ben M'hidi, mirrors Mathieu's answer when he is asked why the FLN hides bombs and weapons in women's baskets. "Give us your helicopter gunships and we will give you our baskets," he tells the reporters who question him. The FLN's war could not be fought with honor. Honor is a luxury that the weak can ill afford when fighting the strong. To the extent that glory is at all related to or awarded for how one fights, it will often fall by the wayside during asymmetrical conflict. On this argument, asymmetrical warfare is particularly "dirty," and there is, by definition, less glory to be had in fighting a dirty war.[26]

3.1.4 Glory and the "Band of Brothers" Argument

Another variety of skepticism about martial glory is focused on the "real" motivation of soldiers to fight. According to this line of argument, soldiers fight because they feel accountable and committed to each other and not for "abstract" goods such as glory, patriotism, or hatred of the enemy.[27] A recent study even suggests that the bonds forged in combat are perceived by soldiers as stronger and carrying more obligations than family ties.[28] The ubiquity of this view of military motivation circumscribes the significance of glory. If soldiers fight primarily to protect each other – if their main commitments are to the other members of their unit, whom they view as family members, martial glory fades as a significant drive. In fact, fighting primarily for glory is often seen by soldiers as a problematic

[25] Gillo Pontecorvo, *The Battle of Algiers* (Igor Film and Casbah Film, 1966). The press conference featuring Mathieu's comment is available here: www.youtube.com/watch?v=eSi8ie8n-6E, and the scene featuring Ben M'hidi's comment is available here: www.youtube.com/watch?v=x5Fdi3Dy69Q.

[26] Asymmetrical war is not new of course; what is relatively new is its ubiquity. For an excellent discussion, see Martin Van Creveld, *The Transformation of War* (New York: Free Press, 1991).

[27] See for example, James McPherson, *For Cause and Comrades: Why Men Fought in the Civil War* (New York: Oxford University Press, 1997) and Stephen Ambrose, *Citizen Soldier: The U.S. Army from the Normandy Beaches to the Bulge to the Surrender of Germany, June 7, 1944–May 7, 1945* (New York: Simon and Schuster, 1997). For a contrary view rejecting the argument that there's one primary reason why soldiers fight, see Christopher Hamner, *Enduring Battle: American Soldiers in Three Wars*, 1776–1945 (Lawrence, KS: University Press of Kansas, 2011).

[28] For an interesting example drawn from the 2011 conflict in Libya, see Harvey Whitehouse, Brian McQuinn, Michael Buhrmester and William B Swann Jr, "Brothers in Arms: Libyan Revolutionaries Bond Like Family," *Proceedings of the National Academy of Sciences of the United States of America*, 111(50) (2014): 17783–85.

disposition that endangers them as well as the mission. The "glory seeker" is the last person one wants to fight alongside because one would not feel safe or cared for next to them.

3.1.5 Glory and Killing from a Distance

A final challenge to the idea of glory arises out of the growing significance of unmanned weaponry in contemporary war. Here, the skepticism is not related to the indignities of mechanized killing as it was in the aftermath of World War I, but to the elimination of danger. Take drone operators as an example. These soldiers, so the argument goes, kill from the safety of climate-controlled rooms that are an ocean away from the battlefield. The diminishment of physical danger shrinks the degree of glory that can attach to their actions. There can be no glory without courage, there can be no courage without risking one's life, and drones allow operators to avoid risks.

That the picture is more nuanced than this, that drone missions involve serious psychological costs for pilots, that they suffer from high rates of PTSD in spite, or perhaps even because, of not being on the battlefield,[29] that they have to deal with jarring transitions from combat operations to everyday life, does not matter to those who make this argument. Unsurprisingly, a RAND report on the attitudes of drone operators found that: "[T]hose in the RPA (Remotely Piloted Aircraft) community feel like the 'red-headed stepchild' of the Air Force ... misunderstood, marginalized, or misutilized by the Air Force and/or supported units. [Drone pilots described] a clear lack of recognition of their efforts by leadership elsewhere in the Air Force and by their peers in other rated career fields. They commented that others believe their job is just like playing video games and that many are unaware of how much skill is required to fly RPAs."[30]

3.2 Running Together Two Kinds of Glory

We've surveyed some of the key sources of skepticism about the idea of military glory. But in spite of all of these objections, and though scholars

[29] See James Dao, "Drone Pilots Are Found to Get Stress Disorders Much as Those in Combat Do," *The New York Times* (February 22, 2013). www.nytimes.com/2013/02/23/us/drone-pilots-found-to-get-stress-disorders-much-as-those-in-combat-do.html, and Eyal Press, "The Wounds of the Drone Warrior," *The New York Times* (June 13, 2018). www.nytimes.com/2018/06/13/magazine/veterans-ptsd-drone-warrior-wounds.html.

[30] www.rand.org/content/dam/rand/pubs/research_reports/RR1700/RR1756/RAND_RR1756.pdf.

don't write about it much anymore, the idea has not lost its appeal. Wherever one looks, promises of glory accompany calls to fight. ISIS propaganda is replete with fantasies of recreating the glory of the lost caliphate.[31] Alt-right talking points promise everlasting glory to those who will defend white Christians from the incursions of multicultural liberal elites. Jimmy Matchette, a member of the Hammerskin neo-Nazi gang wrote to his colleagues from prison that "Being a Hammerskin is the distinct feeling of being set apart from the entire planet. And of knowing we will conquer & overcome all obstacles to achieve our goals and accomplish our great work, knowing that if we fail, all is lost forever and the west will perish. Even though I am locked down in a maximum security federal penitentiary, I wouldn't [have] traded the opportunity for all the gold in the world. You, my true comrades, hold all the glory of victory at your fingertips. We really are the most notorious white power Skinhead group in the entire World!"[32] And then, a world away, there is the rhetoric we are familiar with from our military recruitment videos. The voiceover in a US Marines' recruitment ad aired during the 2018 Super Bowl tells us, over dramatic images of war in air, land, and sea: "[I]t's not just the ships, the armor, or the aircraft; it's something more. It's the will to fight and determination to win found inside each and every marine that answers a nation's call."[33] The rhetoric of "answering a nation's call" could have been taken straight out of Pericles' funeral oration. To be clear: I'm not suggesting that there's an equivalence between these causes. There isn't. Rather, I'm pointing out that well into the twenty-first century, in spite of long-standing, intense skepticism about the term's coherence and value, the idea of military glory is alive and well. How are we to think about the fact that while there are many cultural and technological factors which call this idea into question, it continues, nevertheless, to energize and attract?

Let's recall the parsing we offered between Achillean and Periclean glory in Chapter 2. The distinction between the two may help us solve this puzzle. On the Achillean account, glory is the recognition that comes from fighting well – it is awarded for the combination of physical prowess, skill,

[31] For useful overviews, see this Brookings report on ISIS propaganda: www.brookings.edu/blog/markaz/2015/05/27/why-isis-propaganda-has-been-so-successful-and-how-to-counter-it/ as well as this report from the Tony Blair Institute for Global Change: https://institute.global/sites/default/files/inline-files/IGC_Inside%20Jihadi%20Mind_18.08.17.pdf.

[32] www.adl.org/education/resources/profiles/hammerskin-nation.

[33] www.marinecorpstimes.com/news/your-marine-corps/2018/02/03/marines-seek-young-tough-recruits-in-super-bowl-ad/.

and the ability to face down danger. Achillean glory is amoral – it celebrates the ability to charge into rather than away from physical harm and to function under great duress. Periclean glory, in contrast, is fundamentally moral – it is awarded for defending a worthy cause. It focuses not on a fighter's skill in doing so but on their willingness to sacrifice, and it is that willingness that bestows the rewards: a whitewashing of one's misdeeds and a form of immortality.

This distinction between the two concepts of military glory should be helpful in understanding why the idea persists in spite of the skepticism it attracts. Achillean glory is largely undermined by Owen's criticism. It is, indeed, grotesque to claim that those who are mowed down indiscriminately, that those who "die as cattle" or as "cannon fodder" win glory on the basis of their personal combat prowess. Kicking a soccer ball into machine-gun fire, or mounting a cavalry charge against tanks is not glorious. More than anything, these practices seem, in retrospect, heartbreaking. They yield not great honor but insight into the changing face of war. Indeed, such activities rarely survive once combatants really understand and take the technological changes on board (nobody insists on the cavalry charge anymore; if anyone did they would be crazy, not glorious).

By contrast, the Periclean concept of glory may well survive Owen's criticism. That *how* soldiers die has become more grotesque with the development of military technology does not render their cause or the deaths that uphold it meaningless. It is crude, almost sacrilegious, to submit poetry to this kind of criticism, but here it goes anyway: Owen's statement about dying from gas in World War I features a glaring non sequitur. That choking on one's own lungs after being gassed is awful and unbearable to behold does not mean that "dolce et decorum est" is any kind of lie. The deaths can be meaningful and deserving of glory even if the dying is particularly terrible. WH Auden seems to have understood this, and in his usual mix of decency and common sense, answered Owen's old complaint. Consider this, from "Here War is Simple":

> Here war is simple like a monument:
> A telephone is speaking to a man;
> Flags on a map assert that troops were sent;
> A boy brings milk in bowls. There is a plan
> For living men in terror of their lives,
> Who thirst at nine who were to thirst at noon,
> And can be lost and are, and miss their wives,
> And, unlike an idea, can die too soon.
> But ideas can be true although men die,

And we can watch a thousand faces
Made active by one lie:
And maps can really point to places
Where life is evil now:
Nanking. Dachau.[34]

In spite of the wrenching scenes of men dying thirsty, lost, missing home – the ideas they die to protect may still be valuable: "[I]deas can be true although men die." The cause for which soldiers give their lives can be worth dying for even if the fashion in which they die is abhorrent. The wrongs they seek to repair might desperately need repairing ("maps can really point to places Where life is evil now: Nanking. Dachau") even if repairing them will drudge up all of war's ugliness. In other words, if Pericles was right that dying for a good cause is worthy of military glory, that kind of glory should not be denied because of the terrible conditions of war. Owen's criticism dislodges amoral, skill-based, personal glory because in mechanized, industrial warfare personal skill doesn't mean as much as it once did. But Owen's criticism cannot destabilize Pericles' political version of the term.

Political glory can, of course, be abused. Leaders may try to sell soldiers on going to war by postulating a false cause (the run-up to the 2003 war in Iraq is a striking recent example, a war sold at various times as an effort to punish a dangerous dictator for collaborating with Al Qaeda, an effort to deny that same dictator of weapons of mass destruction, and as a project of spreading peace by spreading democracy). But the fact that political glory can be abused – the fact that the causes we fight for may turn out to be false or a front for other, less worthy aims – does not make the very idea of political glory incoherent; it just makes it suspect and worthy of careful scrutiny.

What about the worries regarding asymmetrical warfare? Can Achillean glory survive them? As a reminder, the concern was that the ugliness characteristic of asymmetrical war – where one side typically takes cover among civilians in order to have a fighting chance and the other pursues its enemies among civilians – diminishes the chances of fighting "honorably" or "well," making atrocities far more likely and, as a result, rendering claims for glory irrelevant.

To the extent that there is a great deal of personal skill involved in prosecuting these wars, Achillean glory does survive the criticism. On the part of the weak, guerillas and particularly guerilla leaders often exhibit

[34] www.poeticous.com/w-h-auden/here-war-is-simple.

physical skill, cunning, and sophistication – in fact, these are usually required for inferior forces to effectively harass stronger ones. Insofar as fighters for the weak party are good at using these skills, and insofar as they take risks or face danger in the process (such as evading capture), such guerillas can properly be the subject of Achillean glory (and, indeed, a glorious local lore does usually build around leaders of guerilla operations – from Che Guevara to Hamas' "Engineer," Yahya Ayyash). However, the foot soldiers of guerilla wars – one thinks for example of those recruited for suicide missions – are often lacking in military skills (indeed they are considered dispensable by the guerilla leaders because they are lacking in such skills). For these people, Achillean glory would not be available.

So much for the weaker parties to such conflicts. Can Achillean glory be bestowed on the stronger side in an asymmetrical war? Insofar as the military activities of the strong involve a similar combination of skill and risk, they too are "eligible" for Achillean glory. Structurally speaking, the strong take fewer risks in these wars and suffer diminished casualties. Consequently, they will come in for less glory. There are, of course, exceptions to this tendency; consider undercover units that are meant to blend into civilian populations in order to extract intelligence and carry out special operations. These activities involve a degree of skill and risk that would seem to partially parallel those taken by the guerillas (in fact these are guerilla operations carried out by the strong against the weak), and as a result, would be rewarded with similar recognition.

How does Periclean glory fair with the complaint from asymmetrical war? It seems like Periclean glory is impenetrable to it, for the same reasons it deflected Owen's criticism. Asymmetrical warfare may be particularly ugly, but the nature of the fighting does not eliminate the possibility of glory. Guerillas, for example, regularly glorify those they send to carry out suicide attacks; the glory attaches not to their prowess in fighting – as we just noted, suicide attackers need not have any such prowess – but to their willingness to sacrifice themselves for the cause. The question of whether Periclean glory attaches to the stronger side in an asymmetrical war is more complicated, as the risks taken by such stronger parties are reduced and there is often controversy about the value of their causes (for example, the French ultimately came around to doubt the importance of keeping Algeria, and some Israelis are torn about whether they should be occupying the West Bank). And yet, if the cause of the stronger party is taken to be of great moment, those who take real risks or die for it are likely to be glorified. It is certainly possible for the cause of the stronger party to be important, and it is certainly possible for those who fight on behalf of that cause to take risks (though on average they take fewer risks than the weak). Stated differently, as with Owen's criticism, the particular ugliness of asymmetrical warfare says nothing about the cause for which one fights. Thus, in principle, both sides to an asymmetrical war can win Periclean glory.

Let's continue to the concern rooted in what soldiers do and don't fight for. As a reminder, the worry was that in real life soldiers fight for their comrades and not for glory, and that, furthermore, those who do fight for glory are "bad soldiers" who put the unit at risk. But the fact that soldiers primarily fight for each other, that their self-understanding as a "band of brothers" is what binds them together and keeps them in the fight, does not indicate that martial glory should be dismissed. It indicates, as we argued in Chapter 2, that glory is not a primary agent of motivation after fighting has commenced. The promise of glory – either Achillean or Periclean – may well serve as a recruitment tool before soldiers join the fight. Also – and this is striking given the complaint – once they are away from the combat zone or done with fighting, members of those "bands of brothers" often insist on proper recognition or attribution of glory to their fallen comrades (again either Achillean or Periclean). Even if fighting for glory is frowned upon by soldiers, the denial of deserved glory after the fight is not tolerated. In other words, there is no necessary conflict between the significance of comradery and the possibility of military glory. The two simply fulfill different social functions during different times.

Finally, it seems like neither variety of glory is consistent with the growing prevalence of remotely-operated weapons. Whether in its Achillean or Periclean permutations, glory involves the assumption of significant physical risk. The disorientation, alienation, and frustration drone operators experience result in psychological harms. But these are not, at least for the moment, the kind of risks that underpin glory. As the social status of psychological harms evolves, this judgment could change. It's worth noting that in the case of drone operations, we may well feel admiration for the technical ability of the drone operator. This admiration is akin to the appreciation that onlookers have when they encounter professional fighting skill – and that appreciation grounds Achillean glory. And yet, as we noted, there can be no such glory here since the element of risk is missing.

3.3 Summary

We began this chapter by reviewing some key critiques of the idea of glory. These include the classical account that finds cultures of honor and glory to be deadly – a set of theories which also explain the early rise of capitalism as an attempt to swap the pursuit of fame with the (safer) pursuit of money. We also reviewed the argument that became influential in the wake of World War I, that it is grotesque to speak of glorious fighting or glorious death in the age of industrial warfare. Other critiques of glory touched on the nature of asymmetrical warfare, the actual attitudes soldiers display towards each other in battle, and the rise of drone warfare. How is it, then, we asked, that in spite of all of these powerful critiques, the idea of glory still permeates public discourse and the rhetoric of war? We suggested that the key to thinking about this puzzle might be a tendency we have to run together the Achillean and Periclean varieties of the term. The Periclean conception of glory seems more robust than the Achillean one as the former withstands important criticisms that the latter collapses under. Since Achillean glory is predicated on individual performance and prowess, it is susceptible to claims that the very nature of modern war often renders individual ability irrelevant (or less relevant than it used to be). In contrast, since Periclean glory turns on the cause combatants fight for, (and not on the manner in which they fight or die) it is immune to these concerns. Achillean glory celebrates skill and excellence. While there are still opportunities to display these in combat, they are certainly narrowing. Periclean glory is awarded on the basis of the willingness to die for something worthwhile. It denotes not excellence but self-sacrifice. There are still plentiful chances for self-sacrifice, even in modern war, and if the cause of war is deemed worthy, the value of such self-sacrifice will survive the ugliness of the dying.

PART II

Humiliation

CHAPTER 4

What is Humiliation?

Recruitment materials for the Islamic State (ISIS) are rife with references to the humiliations Sunnis have suffered at the hands of Shiites and Western powers. Video footage taken by a Vice News team that was allowed to accompany ISIS militants in 2014 depicts fighters for the organization destroying the Sykes-Picot era border between Syria and Iraq – long a symbol of humiliation of Arabs by the World War I colonial powers.[1]

Years earlier, reflecting on the surprise 1973 Egyptian-Syrian attack on Israel, Egypt's President Anwar Sadat explained that the motivation for the Yom Kippur War (or, as the Egyptians call it, the October War) was to reverse the humiliation Israel inflicted on Arab armies in the Six-Day War of 1967.[2]

Sweeping northeast over the map to Russia, much of that country's aggressive foreign policy in recent years – from the annexation of Crimea and the later full-scale invasion of Ukraine to President Putin's insistence on playing a key role in Syria – is framed as a response to Russia's relegation to second-rate power status after the collapse of the Soviet Union and the supposed dismissal of its interests by the West.[3]

China's energetic rise in international relations and its ambitious recent economic and infrastructure projects, including the launch of the Asian Infrastructure Investment Bank, the One Belt One Road initiative, and the push to build and control many of the world's 5G networks, are best understood through (and have been justified by) a desire to emerge from the "century of humiliation" – the long period during which China's interests and its honor were compromised by foreigners. The Chinese Communist Party (CCP) explains and "sells" these massive projects to

[1] www.youtube.com/watch?v=TxX_THjtXOw. [2] Sadat, *Search of Identity*.

[3] See, generally: Lo Bobo, *Russia and the New World Disorder* (London, UK: Brookings Institution Press, 2015), and Shaun Walker, *The Long Hangover: Putin's New Russia and the Ghosts of the Past* (New York: Oxford University Press, 2019).

its own people as part of a move from political humiliation back to the days of the "Middle Kingdom" when China could claim to be at the heart of world affairs.[4]

Finally, a sense of humiliation has been part of the rhetoric that underlies white nationalism in the United States and Europe. In both cases, political messaging speaks of a once ascendant, proud, nationally and culturally cohesive white culture that is being undone by non-white, non-Christian "invaders" who are depriving whites of income, work, status, and ultimately, their very identity.[5]

As these examples suggest, feelings of humiliation fuel a good deal of international and domestic conflict. And yet, political humiliation has received little philosophical scrutiny.[6] What does it mean to be politically humiliated? What does it feel like to be humiliated? Does the subjective sense of humiliation always deserve normative attention? While there is a rich philosophical literature on guilt and shame, humiliation has not received similar conceptual attention. The literature on the role of humiliation in foreign policy and in escalating conflict is even thinner.

Philosophers like to frame and justify their work by saying that they have found a topic that has, until now, been neglected. That can get tiring – many topics are neglected for good reason. But we forgo philosophizing about political humiliation at our own risk. It is a feeling that kills. It is a feeling that is key to understanding the conflagrations of violence all around us – from Ukraine to the Middle East to the American Rust Belt. Failing to think through this feeling is one of those gaps between theory and practice we actually should worry about – so that we can better

[4] See, generally: Eric Anderson, *China Restored: The Middle Kingdom Looks to 2020 and Beyond* (Santa Barbara, CA: Praeger, 2010); Ye Zicheng, *Inside China's Grand Strategy: The Perspective from the Peoples' Republic*, trans. Levine and Liu (Lexington, KY: The University Press of Kentucky, 2011); and Wang Zheng, *Never Forget National Humiliation: Historical Memory in Chinese Politics and Foreign Relations* (New York: Columbia University Press, 2014).

[5] See, generally: Francis Fukuyama, *Identity: The Demand for Dignity and the Politics of Resentment* (New York: Farrar, Straus and Giroux, 2018).

[6] Existing works are primarily in the fields of history, social psychology, and political science. For an outstanding recent historical survey of the role of humiliation in international affairs, see Ute Frevert, *The Politics of Humiliation* (New York: Oxford University Press, 2020). Evelin Linder's important works in the psychology of humiliation are also worth noting: Evelin Linder, *Making Enemies: Humiliation and International Conflict* (Santa Barbara, CA: Praeger, 2006) and Evelin Linder, *Honor, Humiliation, and Terror: An Explosive Mix – And How We Can Diffuse It with Dignity* (Dörzbach, Germany: Dignity Press, 2017). Finally, for further recent treatments, see Deepak Tripathi, *Imperial Designs: War, Humiliation & the Making of History* (Washington, D.C.: Potomac Books, 2013); Bertrand Badie, *Humiliation in International Relations: A Pathology of Contemporary International Systems* (Portland, OR: Hart Publishing, 2017); and Joslyn Barnhart, *The Consequences of Humiliation: Anger and Status in World Politics* (Ithaca, NY: Cornell University Press, 2020).

understand what drives nations against each other, when to heed proclaimed grievances, and how to avoid aggrieving in the first place. In this chapter and the next, I describe some of the conceptual elements of political humiliation and, then, focus in on the meaning of humiliation in international affairs.

4.1 Humiliation as a Failure of Recognition

The few philosophers who take up the meaning of political humiliation tend to describe it as a failure of recognition. Following Ernst Bloch, George Herbert Mead, and, of course, Hegel, Axel Honneth maintains that the development of personal integrity – a healthy sense of one's self and one's possibilities in the world, requires recognition: An affirmation and reflection of one's abilities, potential, achievements, and social standing. "Human individuation," he writes, "is a process in which the individual can unfold a practical identity to the extent that he is capable of reassuring himself of recognition by a growing circle of partners to communications. Subjects capable of language and action are constituted as individuals solely by learning, from the perspective of others who offer approval, to relate to themselves as beings who possess certain positive qualities and abilities."[7]

When such recognition is denied, a series of negative psychological consequences follow. Honneth describes these failures of recognition as injured dignity or disrespect, and the psychological ramifications include a sense of shame, insult, indignation, and humiliation. He does not spend much time carefully distinguishing between these responses, and focuses, instead, on classifying the kinds of slights or failures of recognition that lead to them. The most severe failure of recognition is one that involves the negation of another's control over their body – through violent acts like torture or rape. These acts deny recognition of bodily autonomy and this is devastating in part because such recognition is necessary to develop a secure, healthy sense of embodiment. Next comes systematic political exclusions and dismissals – the failure to recognize others as bearers of rights and as subjects worthy of the law's consideration. The extreme example here is slavery but the depredations of institutionalized racism (Jim Crow, apartheid) fall under this category as well. Finally, there is the dismissal of one's way of life – a rejection of an individual's or group's

7 Axel Honneth, "Integrity and Disrespect: Principles of a Conception of Morality Based on the Theory of Recognition," *Political Theory* 20(2) (1992): 189.

sexual preferences, aesthetic tastes, lifestyle choices, and cultural achievements. Examples here range from contempt of atheism and atheists to disgust with gay culture.

These denials of recognition are detrimental not only to our sense of self and our ability to develop as competent moral agents. With Ernst Bloch, Honneth argues that the sense of indignation and injury suffered in these cases fuels social upheavals as well. Large social revolts draw not only (perhaps even not primarily) on economic or political injustices but, rather, on the sense of indignation, rejection, insult, and humiliation created by the failures of recognition inherent in these injustices. "Were it not for the added feeling of wounded dignity," Honneth maintains, "... the mere experience of economic distress and political dependence would have never become a driving force of the practical revolutionary movements in history."[8]

Humiliation for Honneth is, then, part of the range of harmful results of misrecognition – a psychological cost inevitably born by creatures who cannot mature without the acknowledgment of others and who suffer, in a variety of ways, when that acknowledgment is withheld.

The most comprehensive work of political philosophy on humiliation is, by far, Avishai Margalit's *The Decent Society*.[9] Margalit's account focuses on humiliation in domestic politics. A decent society, he argues, is one that does not humiliate or condone humiliation; it is a polity in which institutions don't humiliate the citizens and the citizens don't humiliate each other – at least not with the tacit or explicit encouragement of the state. Margalit understands humiliation as a subset of cruelty and he understands cruelty (following Judith Shklar) as the "ultimate evil." In particular, humiliation is the psychological aspect of cruelty, its "extension ... from the physical to the psychological realm of suffering."[10] We can be humiliated because we can be hurt not just physically but also by symbolic interactions. Since preventing cruelty "is the supreme moral commandment,"[11] understanding and preventing political humiliation is more important than understanding and realizing the requirements of a Rawlsian *just* society. It is, in other words, more urgent to eliminate humiliation from the public sphere than to work towards a rights-backed social order; it is also practically easier to identify and counteract humiliating acts than to identify and encourage respectful ones. Margalit's account

[8] Ibid., p. 196.
[9] Avishai Margalit, *The Decent Society* (Cambridge, MA: Harvard University Press, 1996).
[10] Ibid., p. 85. [11] Ibid., p. 87.

of the decent society is, then, negative: He is interested in the wrongs a society must avoid in order to clear the normatively modest bar of non-humiliation.

Margalit defines humiliation as "behavior or a condition that constitutes a sound reason for a person to consider his or her self-respect injured."[12] Humiliation is understood primarily as the denial or negation of self-respect, as a privation rather than as a well-articulated, independent disposition. The relevant concept of self-respect to posit against humiliation is one focused on intrinsic human dignity rather than the social honor of, for example, receiving medals and commendations.[13] We can be humiliated by other individuals (as in cases of bullying or domestic abuse), and we can be humiliated by institutions (as in the skewing of the criminal justice system in the United States against young men of color).[14]

As the definition makes clear, this theory of humiliation is normative rather than psychological. There are right and wrong grounds for feeling humiliated, and the simple presence of the feeling is not decisive as far as Margalit is concerned. People can feel humiliated when they ought not to and they can fail to feel humiliated even when, by rights, they should. Furthermore, only behavior or conditions created by people can be humiliating, natural circumstances cannot. One can't be humiliated if one gets hurt by a devastating earthquake that reduces their home to rubble, though one can be humiliated by the failure of authorities to help and by the indifference of their neighbors. To take a recent example, Puerto Ricans cannot, properly speaking, be humiliated by the degradation of their life conditions resulting from the hurricane of 2017. They can be humiliated, however, by the US government's failure to respond to their plight adequately. Humiliation, then, is a social vice and a social condition. We are properly humiliated only by the consequences of bad behavior (or culpable omissions).

In his account, Margalit pushes back against two views about humiliation which he considers extreme: He dismisses the anarchist who holds that all institutions are, by definition, humiliating. This position is implausible because the anarchist offers no account of how a dignity-sustaining standard of living can be reached without institutions. He also rejects the Stoic's insistence that a truly free, self-respecting person could never be humiliated by another person or institution. Margalit argues that self-respect cannot take hold in a social vacuum – it is, rather, part of a social practice. Further, Margalit is unimpressed by the authenticity of the

[12] Ibid., p. 9. [13] Ibid., p. 44. [14] Ibid., pp. 128–29.

Stoic response to adversity. Rather, he views that response as a "defense mechanism" that is a form of self-imposed discipline because a desired form of recognition cannot be had.[15] Supposedly Stoic strategies such as the appropriation of derogatory terms used by the aggressor can mitigate but do not eradicate humiliation.

Margalit argues that humiliation has three forms – first and foremost, it consists in removing people from the human community (as when we demonize them or liken them to animals – the history of demonic references to Jews and animalistic references to African Americans come to mind), but it also manifests in the negation of control (as in torture), and in simply ignoring or looking through others. The distinctions here are not, of course, clear-cut. The denial of control over one's body can be understood as a special case of removing someone from the human community if we follow the view that self-control and self-direction are definitive of what it means to be human. On this view, which takes freedom as ontological to humans, treating them as if they are not free is treating them like nonhumans.[16]

When humiliation takes the form of overlooking others, an assailant does not dehumanize or negate control but, rather, fails to see someone fully or precisely; this is perhaps how an aristocrat sees a servant or how an oblivious senior executive sees a parking attendant. It is not clear if Margalit sees this third form of humiliation, which he discussed in less detail, as more benign than the first two.

Crucially, Margalit tells us that humiliation involves seeing others *as if* they were nonhumans. All dehumanization is, in the end, pseudo dehumanization because it's almost impossible for us to really see humans as nonhumans.[17] But what specifically does it mean to see (or fail to see) another as human? Margalit writes: "[S]eeing a human being as human means . . . seeing the human body and its parts in the mental terms they non-literally exemplify." We see others as human when we look at them and notice the dispositions and moods their bodies (facial expressions, body posture) convey. The ability to interpret the meaning of physical composure and gestures depends on some social knowledge and empathy. But seeing humans in their human aspect is not really a choice once such rudimentary social knowledge and capacity for empathy are in place: "I see

15 Ibid., pp. 24–27. 16 Ibid., pp. 116, 118.

17 Ibid., pp. 91–92. Here, Margalit foreshadows recent historical, philosophical, and anthropological discussions claiming that mass violence does not involve literal dehumanization. For an excellent overview, see www.newyorker.com/magazine/2017/11/27/the-root-of-all-cruelty.

your eyes as mocking and your hands as nervous just as I see your eyes as brown and your hands as sinuous . . . but just as I see your eyes as mocking and your hands as nervous, I see you as human and I cannot see you otherwise."[18] When we see Gorbachev's sad face during his resignation speech, it is a different act of observation than seeing his famous birthmark, but it is not more voluntary.[19] Under normal conditions, one notices that his interlocutor is sad. She does not intentionally "turn on" the capacity to do so. Seeing humans as human, then, means "seeing them as tagged with the labels of human psychology,"[20] but that doesn't mean that these attributes are less real than the physical ones.

Of course, some of us may be blind to the "human aspect" in others; some of us will fail to perceive the emotional meaning of physical signs. But for Margalit, not seeing someone you look at as human – namely under the label of human psychology – is a hardware malfunction or a pathology. In other words, a true failure to see someone as human constitutes not humiliation, but a cognitive impairment similar to the one afflicting Oliver Sack's famous patient who mistook his wife for a hat.[21] Humiliation involves training or at least urging oneself to ignore the human in others or discount it – but it is a training away from the normal mode of perception and a residue of the original way of seeing always lingers.

Put differently, while humiliation often involves the exclusion of another from the human community (seeing her as a beast, a vermin), this still takes place while knowing – at varying levels of clarity – that those so perceived are actually human. Drawing on Hegel's Master-Slave Dialectic, Margalit notes that humiliation must involve a paradox: For it to work as an act of true degradation it requires the trampling of someone whose humanity is at least tacitly acknowledged – "[T]he victim must be taken as someone with awareness, thus possessing implicit human worth, in order for the act of humiliation which denies his humanity to take place. Humiliation is intended both to prove absolute superiority and to win recognition, which is a conceptual impossibility. Absolute superiority can be achieved only with regards to what is not human; recognition can be attained only from other humans."[22]

An important part of what it means to be rejected from the human community is to suffer the rejection of one's legitimate encompassing group. Encompassing groups are the different collective affiliations that

[18] Margalit, *The Decent Society*, p. 95. [19] Ibid., pp. 94–95. [20] Ibid., p. 95.
[21] Ibid., p. 101. [22] Ibid., p. 110.

layer a life with meaning: being Jewish, Muslim, Catholic, Irish, a Free Mason, and so on. Different encompassing groups prescribe different ways of being human. A rejection of these dismisses or negates one's way of being human and thus constitutes humiliation.[23] A society that through its policies or institutions makes one ashamed of an aspect of her group identity is a humiliating society and, thus, an indecent one (making one ashamed of being Catholic or Jewish through the use of derogatory stereotypes can serve as a useful example). This said, the aspect of identity that one is made to feel ashamed of has to be legitimate. One can't be humiliated by dismissal and derision directed at her belonging to an organized crime collective. A society that makes a citizen ashamed of such an illegitimate affiliation or aspect of her identity is not made indecent by that.

A final form of political humiliation involves rejection from full citizenship. There can be no second-class citizens in a decent society: Having citizens with curtailed rights removes them from full social participation and, thus, from the community of responsible adults (who deserve full-fledged rights). Second-class citizenship is humiliating, then, both because it excludes its victims from full legal personhood and because it involves the forced determination of one's fate and status by others. On this account, discrimination is not merely a matter of distributive justice; it is, first and foremost, humiliating as it signals that certain citizens are not full-blown moral agents worthy of the law's consideration.[24]

4.2 Challenges for the Misrecognition Theory of Humiliation

Seminal as Margalit's and Honneth's accounts are, they leave us with significant lacunae. In what follows, I further develop the misrecognition account of humiliation by thinking through its contagious nature. More radically, I question the necessity of the recognition-humiliation connection underlying Margalit's and Honneth's discussions. Both philosophers assume that humiliation is the flip side of recognition. Both provide various examples of such misrecognition (rejection of a legitimate encompassing group, looking through others, torture). I pushback against their assumption by exploring whether it is possible to feel humiliated and recognized at the same time and whether it is possible to be humiliated by those whose recognition does not matter to us.

[23] Ibid., p. 143. [24] Ibid., p. 154.

4.2.1 Contagion

ISIS has been successful at convincing young Muslims in the West that they are part of a unified Muslim nation or *umma* and that affronts to Muslim dignity – even old ones – constitute current injuries to Muslims everywhere. Why has this effort worked so well, especially in the case of people who have not suffered serious personal humiliation?[25] How, for example, did a young Muslim engineering student from London become personally humiliated, in 2016, by the 1916 Sykes-Picot agreement which summarily dismissed promises to Arab leaders fighting on behalf of the British and French in World War I? That old agreement features prominently in ISIS propaganda. Why is it such a powerful material? Margalit notes, in passing, that humiliation can be contagious: "[I]t is an emotion we may feel as a result of mere identification with others, even if we are not the direct victims of the humiliating behavior. If we identify with the victim in that we share the characteristic for which he is being humiliated, then we also have a justified reason for feeling ourselves humiliated."[26] One assumes (though Margalit does not explicitly argue) that this is so because the sense of belonging to an encompassing group is formative of our identity – so much so that we are harmed if the encompassing group is humiliated.

Margalit is certainly on to something: Humiliation and the visceral response to it are contagious – across time and space. From Western ISIS recruits who were moved by the plight of Palestinians to young Armenians inflamed by the World War I era Armenian Genocide, there is no doubt that, as a psychological fact, the sense of humiliation travels. But there are important normative questions here that Margalit does not ask. He argues that as long as the encompassing group suffering humiliation is legitimate (one can't be humiliated on behalf of injuries sustained by her branch of the Mafia), a sense of humiliation by an individual affiliated with that group is legitimate. But, surely, that is not enough. What if, as in the case of ISIS, the encompassing group that has purportedly suffered humiliation (the *umma*) is legitimate, but the actor making the claim for humiliation on its behalf (ISIS) is not? While there is a history of humiliation of Muslims by Western powers (and the Sykes-Picot agreement is part of that history), should ISIS, with its nihilist, genocidal

[25] For an interesting account of well-off young Britons joining ISIS, see www.economist.com/middle-east-and-africa/2014/09/01/it-aint-half-hot-here-mum.

[26] Margalit, *The Decent Society*, p. 32.

interpretation of what it means to be a Muslim be allowed to be the arbiter of what counts as a slight? It seems, then, that in order for a claim of "distant" humiliation to be legitimate, both the encompassing group and the person making the claim on its behalf must be legitimate.

But the questions do not stop here. How does distance from the injury – in time and space – impact the legitimacy of a claim of humiliation? Should an injury in the distant past gradually lose some of its moral force to ground a claim of humiliation? When, for example, does the Sykes-Picot betrayal begin to lose its bite? When do the ravages of the crusades fade? When, in short, does a sense of justified historical grievance turn into a worrying inability to let go of the past?[27] Does an injury to geographically remote members of a group have less of a claim than an insult to closer ones?[28] Does an injury that has been redressed – by reparations, an apology, or other forms of acknowledgment, lose some of its ability to humiliate? Can I, in other words, still legitimately feel humiliated on behalf of my group, if the injuries against it have been recognized in a morally satisfactory fashion? Each of these concerns could occupy its own chapter. For our purposes, it is sufficient if they establish that the question of the contagious nature of political humiliation is far more complicated than Margalit allows for.

4.2.2 *Questioning the Relationship between Humiliation and Recognition*

Both Honneth and Margalit understand humiliation as the flip side of recognition. This view suggests that humiliation is primarily a derivative disposition – one that comes about when various manifestations of its positive counterpart (respect/recognition) are withheld. But is that really so? Is there, indeed, a full symmetry here? If we can be humiliated by those we do not recognize, by those whose recognition we do not seek (or actively reject), and conversely, if we can feel humiliated in spite of being recognized, then there are reasons to doubt the dependency. Let's explore these possibilities.

[27] If, as Jeremy Waldron argues, in some cases historical injustice can be superseded by the passage of time, is the same true of the salience of group humiliation? See Jeremy Waldron, "Superseding Historical Injustice," *Ethics* 103(1) (1992): 4–28.

[28] Here one thinks of something like the Talmudic principle stating that "the poor of your own city take precedence" (Bava Metzia 71a: 3–4). Translation available here: www.sefaria.org/sheets/216801.4?lang=bi&p2=Bava_Metzia.71a.3&lang2=bi.

It is certainly possible to feel decently treated by those whose authority over us we completely reject. Some of the Nazi defendants at Nuremberg, for example, reported that the tribunal treated them decently and humanely, even though they completely rejected its authority to sit in judgment of them.[29] But if that's possible, shouldn't it also be possible to feel humiliated by those whose recognition we do not seek – by those whose recognition one actively rejects? ISIS jurisprudence certainly rejects the very possibility of accepting the rights of apostates or *kuffar* (secular Sunnis, Shiites, non-Muslims); these groups cannot be afforded any recognition and, in dealing with them, it is impermissible to reach anything more ambitious than a temporary truce. And yet, ISIS speakers insist that Muslims have been humiliated, for centuries, by such actors. But if that's the case the source of humiliation cannot be a failure of recognition, as *kuffar* don't have the status to grant or withhold recognition to true proponents of the faith. It follows that ISIS is either mislabeling the emotional response they call "humiliation" or, otherwise, that it is possible to be humiliated by those you do not recognize.

Consider another, semi-fictionalized example, drawn from the recent television series *Turn*. The show's topic is counterintelligence warfare during the American Revolutionary War (it is based on real events and draws on the book *Washington's Spies* which documents a rebel spy ring).[30] Caleb Brewster is a rebel spy masquerading as a profiteer selling goods to the Patriots as well as the British. He is caught by a British intelligence officer, John Graves Simcoe, and severely tortured and abused. Brewster eventually makes it out of captivity. He does not remember if he gave up any secrets when he was tortured. Though he does not recognize Simcoe as a worthy adversary, though he does not recognize Simcoe's authority to hold him captive or try to extract information from him, though he feels nothing but profound contempt for Simcoe, though the two fundamentally view each other as less than human, Brewster feels deeply humiliated after his captivity, so much so that his very sense of agency erodes. To reiterate: If Brewster does not think that Simcoe is fully human, if he does not view Simcoe as an agent capable of extending recognition in the first place, then, based on Honneth's and Margalit's accounts, Brewster could not have been humiliated by Simcoe. But he felt he was. Either

[29] See Gary Bass, *Stay the Hand of Vengeance: The Politics of War Crime Tribunals* (Princeton, NJ: Princeton University Press, 2002).

[30] Alexander Rose, *Washington's Spies: The Story of America's First Spy Ring* (New York: Bantam, 2014).

Brewster was wrong about what he felt or the sense of humiliation can have other sources, unrelated to recognition.[31]

Finally, there is the possibility of feeling recognized and still humiliated. In an alternative world, Simcoe and Brewster might have had great respect for each other as full-blown moral agents, albeit on the opposite sides of a conflict. Simcoe might have had unwavering admiration for Brewster and still tortured him due to circumstances he regretted (urgently needed information). In fact, the Continental Army did hang John Andre, a senior British intelligence officer. Both sides agreed that hanging was a humiliating mode of execution, inappropriate for an officer, and that the dignified alternative would have been to put Andre in front of a firing squad. And yet, for a set of political reasons that we need not get into here, Andre was hung as if he were a petty criminal. The Patriots regretted this. Alexander Hamilton famously wrote of Andre's execution: "[A]mong the extra ordinary circumstances that attended him, in the midst of his enemies, he died universally esteemed and universally regretted."[32] In death, John Andre, Adjutant General of the British Army, was both recognized and humiliated. Either the reported sense of Andre's humiliation (and the sympathy felt by his executioners for that humiliation) was mistaken, or Margalit and Honneth's insistence on neatly tying humiliation and misrecognition is misplaced. Or, if we were to offer an example from a very different and very contemporary context, consider the rationale behind guaranteed basic income programs like the one proposed by Andrew Yang (similar plans are being partially implemented in some European countries). The rationale behind these initiatives is laudatory and predicated on a profound respect for agency: Advances in automation can herald a rise in leisure and provide an opportunity for people to sever the traditional and problematic connection between work and meaning. The state may honestly look to programs that guarantee basic income in order to free people to pursue new interests

[31] Perhaps Margalit would attempt to resolve this puzzle by claiming that Brewster and Simcoe actually pseudo recognized each other and it is this pseudo recognition that is at the heart of the ability to feel humiliation. The trouble with this explanation is that both agents vehemently reject it. It flies in the face of their experience and beliefs. It is worthwhile asking whether there's an explanation of Brewster's sense of humiliation that accounts for rather than rejects his description of events. Or perhaps Margalit and Honneth would argue that Brewster's sense of humiliation was caused by the negation of agency that is entailed by torture – by the failure to respect Brewster's autonomy over his body. But again, at least in the televised account, Brewster seems to feel humiliated because he no longer sees himself as someone capable of withstanding torture – as a tough, impenetrable, sturdy guardian of secrets. The classical recognition-based account does not do justice to this experience.

[32] Letter from Alexander Hamilton to Lieutenant Colonel John Laurens (October 11, 1780). https://founders.archives.gov/documents/Hamilton/01-02-02-0896.

and hobbies. The idea is essentially respectful: To create the freedom for people to reflect and act on their own interests. And yet it is also fully compatible with a genuine feeling of humiliation on the part of people who now no longer work, who lost the sense of professional identity that went along with their work, and who no longer feel self-sufficient – in that, they can't support themselves and their families like they used to. Their subjective sense of dignity was psychologically tied to their professional identity and self-sufficiency, and the loss of those can be catastrophic and legitimately humiliating even if it was not brought about by intentional acts of misrecognition (indeed, even if all the intentions involved trended in the opposite direction).

All of these examples cast doubt on the necessary connection that Honneth and Margalit draw between recognition and humiliation. The discussion suggests that humiliation, rather than being the flip side of recognition, is an independent affective disposition. The discussion also suggests that an anti-humiliation policy that is exclusively focused on recognizing and showing respect for others may fall short of preventing humiliation.

4.3 Humiliation and the Sense of Being Replaced

The possibility of being humiliated without being misrecognized points to a structural lacuna in Margalit's account. His theory is normatively compelling; it gives a keen sense of what's wrong with humiliating. But it tells us little about the experience of being humiliated. Sure, sometimes what it feels like to be humiliated is to be treated like an object. That's how Primo Levi felt when a Kapo wiped his greasy hands on Levi's coat. Levi was struck that, of all the things that could have humiliated him in Auschwitz, this actually did. More prosaically, that's how Gwendolen Harleth felt in George Eliot's *Daniel Deronda* when her cold, estranged husband told her to wear jewelry he knew she hated.[33] But that's not all there is to humiliation. There doesn't need to be a deliberate attack on one's self-respect for one to feel humiliated. Assuming that such a connection is necessary fails to account for the richness of our experience of humiliation.

Margalit could well respond that when that connection is missing, one may *feel* humiliated but one would lack the normative grounds for feeling that way. In other words, one would be *wrong* to feel humiliated if there's

[33] For a wonderful discussion of both examples, see Vivian Gornick, "Put on the Diamonds," *Harper's* (October 4, 2021). https://harpers.org/archive/2021/10/put-on-the-diamonds-notes-on-humiliation-vivian-gornick/.

no misrecognition. That may be correct as far as it goes. But this response fails to consider that people regularly do feel humiliated for these "wrong" reasons and this sense of humiliation can be powerful and, more importantly, politically dangerous (even if it is normatively groundless).

On one level, it is futile to give an account of what it feels like to be humiliated, because people can be humiliated by a surprising variety of circumstances.

In his essay on torture, Jean Amery reports that people complain about losing their dignity for all kinds of reasons – from not being allowed to bathe to being rudely addressed. [34] And Vivian Gornick notes that "the circumstances that can make people [feel humiliated] are as variable as the persons themselves." She goes on to provide some helpful examples: "I have a cousin, a doctor, who is humiliated if he is shortchanged in the grocery store. His wife, too: if another woman is wearing the same dress at a party . . . I once had a mother-in-law whose critical observations amused me; my husband's next wife felt humiliated to the bone by them."[35]

In short, our sensitivity to humiliation varies: A comment that makes one person laugh, will annoy another, and humiliate a third. What, then, is the point of coming up with a phenomenological account of the feeling? There's only a point if there are some commonalities underlying the differences.

In Chapter 5, I will focus on one such commonality. Let's briefly preview that discussion: Feeling humiliated often boils down to a sense of being replaced. This sense of being replaced is central to political conflict. But it's there in interpersonal relationships as well. The jilted partner is humiliated because he feels replaced by someone else – younger, newer, more exciting. A friend is humiliated when she is replaced by a new companion and becomes irrelevant, almost transparent. Vivian Gornick describes an instance of humiliation she remembers as a young girl. Her closest friend had abruptly turned cold and Vivian could not figure out why. But ultimately: "[I]t registered that whenever I spotted her, in the hall or the schoolyard, she was in the company of a girl new to the school. One day, I approached the two of them in the yard.

"Sheila," I said, my voice quivering, "aren't we best friends anymore?" "No," Sheila said, *her* voice strong and flat. "I'm best friends now with Edna." I stood there, mute and immobilized. A terrible coldness came over me, as though the blood were draining from my body; then, just as swiftly,

[34] Jean Amery, "Torture," in Sidney Rosenfeld and Stella Rosenfeld (trans.), *At the Mind's Limits* (Bloomington, IN: Indiana University Press, 2009): pp. 21–41.

[35] Gornick, Put on the Diamonds.

a rush of heat, and I was feeling bleak, shabby, forlorn, born to be told I wouldn't do, not now, not ever. It was my first taste of humiliation."[36]

Much more dangerously, this sense of being replaced shows itself in domestic and international politics when certain groups report a sense of humiliation from losing their traditional ascendency and sense of uniqueness in favor of others (immigrants, non-Westerners, non-whites, non-Christians, women).

In the United States and abroad, white nationalists and white supremacists are making the explicit claim that they are being replaced by foreigners. The same might be said about extremists in other parts of the world. In the propaganda distributed by the self-proclaimed Islamic State, we find calls to reverse the shame and humiliation that Arab nations have suffered. This humiliation is defined as lost ascendancy – a position of power and influence that Muslims have purportedly lost in recent centuries to "apostates" and "crusaders."

The language of replacement is not limited to terrorist groups. Indeed, national actors often make structurally similar claims. Russia's leaders, for example, have complained about being humiliated by the Americans and their NATO allies. What the humiliation consists in, is a sense of being replaced on the world stage – having lost influence and ascendency as an international power after the end of the Cold War. Similarly, Chinese politicians have talked about their country's "century of humiliation" at the hands of Western nations. Here too we can understand much of the country's foreign policy as an attempt to reverse China's replacement on the world stage – as a project for re-establishing it as the "Middle Kingdom," a global hub of commerce and culture and a leader in the creation of internationally vital infrastructure.

In these examples, as in other cases of humiliation, it is important to distinguish the normative weight of the claim (ISIS leaders and white supremacists have no moral grounds for complaining about being replaced) and its psychological heft. What are we to make of claims of replacement that are morally groundless and yet psychologically salient? We will take this up in the next chapter.

4.4 Humiliation, Shame, and Embarrassment

Let us take stock of what we have learned about political humiliation so far.

[36] Ibid.

Margalit understands humiliation as an act of cruelty, a practice compounded by the fact that humans are capable of unique suffering – experiencing pain not exclusively from physical hurt but also due to refusal of recognition or acknowledgment of key aspects of their moral standing. Humiliation can be levied by institutions on individuals or by individuals on each other (with or without institutional support or passive acquiescence). Humiliation is a personal or social vice – it involves an agent's or an institution's intentional and culpable act. On this view, being subject to a natural disaster can't be humiliating (though the institutional mismanagement of its aftermath can be). Humiliation typically involves one of these three violations of status – removal from the human community or dehumanization (viewing another as an animal, a monster), a negation of control over one's body or denial of autonomy, or contempt for one's way of life. The three conditions can be but are not necessarily distinct, and an act of humiliation will often involve more than one of these features (in fact, one feature will often be justified by another: Seeing someone as less than human will justify taking control of their body and will prompt contempt for their way of life). These varieties of humiliation often manifest themselves in a rejection of the subject's encompassing group – a group that gives detail and weight to the humiliated person's identity, or a refusal to admit that person into full citizenship. So, for example, if X is Jewish, part of what it means to dehumanize X is to make anti-Semitic claims about all Jews. If X is gay, part of what it means to dehumanize X is to refuse to extend full citizenship rights to her (and thus signal that she does not merit the entitlements that normally come with our acknowledgment of personhood).

Humiliation requires that a perpetrator train herself to see the victim as less than human – to ignore human features that they would otherwise be naturally disposed to see. This amounts to teaching oneself to diverge from the normal mode of perception, but a residue of the original way of seeing always lingers.

Adding to Margalit's account, I have argued that this sense of humiliation travels both in space and in time – it is possible to feel humiliation along these lines for things that happened a long time ago or things that happened to other people who are far away, though we have raised some questions about when one's claim on humiliation begins to fade under these circumstances.

I have also argued that there is more to political humiliation than misrecognition or the denial of one's human dignity. Humiliation and recognition are not different sides of the same coin; to be humiliated is not simply to be disrespected in the Kantian sense of that term. After all, we can feel humiliated by those whose recognition we don't seek or care about, and we can feel severely humiliated even when our status as rational members of the human community is fully recognized or when others have no intention of questioning it.

With this initial account of political humiliation crystalized, it is time to further delineate the nature of humiliation by distinguishing it from two related but ultimately quite different dispositions – shame and embarrassment.[37]

Shame arises when others, whose opinions we care about, have perceived us committing a culpable act. Unlike guilt, which is primarily a private disposition, shame typically requires an audience. The degree to which it is felt depends on how wrong the act was, how public it was, and the significance of the audience. Shame can and often does attach to the very character of the agent feeling it, unlike guilt, which is usually more restricted to feelings about a particular deed.[38] Shame, in other words, "infects" the agent's self-perception: When one is ashamed, one feels that one is bad and not simply that one has done something wrong.

Since shame easily attaches to a culpable agent's character, the sentiment is often subject to manipulation by those wishing to harm the agent. Thus, the act of *shaming* consists in trying to convince others to draw conclusions from X's culpable act to X's character, and political shaming consists in drawing conclusions from the culpable actions of one or more political actors to the national character of an entire people. Daniel Goldhagen's book *Hitler's Willing Executioners* is sometimes held up as an example of the latter,[39] as Goldhagen explains the horrors of Nazism with reference to essential attributes in the German national character. Shaming carries obvious psychological risks for the targets. Small wonder that psychologists caution against telling a child that he is a "bad boy" in response to an impermissible act; better, they advise, to focus on censuring specific

37 The account of shame and embarrassment that follows draws on Nir Eisikovits, "Embarrassment and Political Repair," in Thomas Brudholm and Johannes Lang (eds.), *Emotions and Mass Atrocity: Philosophical and Theoretical Explorations* (Cambridge, UK: Cambridge University Press, 2018): pp. 262–76.

38 Roger Petersen, "Guilt, Shame, Balts, Jews," in Daniel Chirot, Gi-Wook Shin and Daniel Sneider (eds.), *Confronting Memories of World War II: European and Asian Legacies* (Seattle, WA: University of Washington Press, 2014): pp. 258–83, p. 263. See also, Alba Montes Sánchez and Dan Zahavi, "Unraveling the Meaning of Survivor Shame," in Thomas Brudholm and Johannes Lang (eds.), *Emotions and Mass Atrocity: Philosophical and Theoretical Explorations* (Cambridge, UK: Cambridge University Press, 2018): pp. 162–84 – "Shame," they write, "persists even when the shame-inducing situation has changed or ceased. It is felt as more painful and unbearable, and is taken to reveal or disclose some overshadowing truth about one's own being" (p. 177).

39 Peterson, Guilt, Shame, Balts, Jews, p. 265.

behavior.[40] Furthermore, as Catherine Lu reminds us in her discussion of the relationship between acts of shaming and feelings of shame, those whom we shame will often not feel what we are trying to distill. Instead, they could lash out against us: "[T]hose who are made to feel ashamed may respond with rage against that which provoked the painful recognition of their inadequacies, rather than engage in self-criticism and reform – this is the dynamic of 'shame–rage spirals'."[41]

In her book *Hiding from Humanity: Disgust, Shame, and the Law*,[42] Martha Nussbaum argues that there is a tension between the use of legal shaming practices (examples include a judge who ordered a defendant convicted of larceny to wear a shirt that identified him as being on probation for theft as well as a requirement that convicted drunk drivers use a bumper sticker denoting them as such) and the role of law as an important institution for protecting human dignity and counteracting stereotypes. Shaming, Nussbaum argues, cannot effectively stigmatize vice, because it is hard to be sure that the behaviors stigmatized really are problematic rather than simply deviant from dominant norms, which may themselves be wrong. Supporters of shaming, Nussbaum writes, "offer no sustained analysis of the operations of shame and stigma that would help convince us that the penalties are likely to be reliably directed at vice rather than at unpopular people or people who make the dominant majority uncomfortable."[43]

On the other side of this debate, we find thinkers who argue that shame can be a valuable and productive social sentiment. True, it can be misused; we can be ashamed of the wrong things (if, for example, we are ashamed of something we are not guilty of) or an attempt at shaming may generate an angry, defensive response. However, the fact that the emotion is subject to manipulation or can be experienced inappropriately does not render it morally and socially worthless. In their novel account, Julien Deonna, Raffaele Rodogno, and Fabrice Teroni have argued that feeling ashamed can reflect someone's considered understanding that she has violated an

[40] See, for example, Paul Gilbert and Coby Gerlsma, "Recall of Shame and Favouritism in Relation to Psychopathology," *British Journal of Clinical Psychology* 38 (1999): 357–73.

[41] Catherine Lu, "Shame, Guilt and Reconciliation After War," *European Journal of Social Theory* 11 (3) (2008): 374.

[42] Martha Nussbaum, *Hiding from Humanity: Disgust, Shame, and the Law* (Princeton, NJ: Princeton University Press, 2004): chapter 3.

[43] Martha Nussbaum, "Danger to Human Dignity: The Revival of Disgust and Shame in the Law," *The Chronicle of Higher Education* (August 6, 2004).

important social and moral norm; the feeling may well lead that person to reevaluate their behavior and mend their ways.[44]

Others note the potentially positive role some forms of shaming can play in the wake of conflict and mass atrocity. John Braithwaite recognizes the possibility of "reintegrative shaming" – which is meant to denote that a generally decent person has done a bad thing for which they must repent. This kind of shaming is still consistent with the possibility of redemption – indeed, it may help bring about reintegration if a perpetrator has taken responsibility for their behavior.[45]

Unlike shame, embarrassment doesn't stem from the sense that one has done wrong or that one has somehow become bad. Its emergence depends on the gap between a valuable self-description/understanding and how one's actions are actually perceived. While shame is often accompanied by regret for having performed the action and comes with a heavy heart for the harm inflicted or the moral standard violated, embarrassment primarily involves regret about having been seen or caught.[46] The contradicted self-description needs to be important, and, like in shame, the audience matters. The more central the challenged self-description is to the identity the subject would like to project, and the more important the audience who witnesses the discrepancy, the greater the potential for embarrassment. This is true for the child who thinks herself "cool" and independent, only to have a parent show up and enthusiastically embrace her in front of friends, and it is true for the nation that views itself as a champion of freedom only to have its ill-treatment of minorities, prisoners, or vulnerable populations aired out. Aristotle tells us that shame is not so much about the consequence of the loss of esteem, but rather about its loss as such. Embarrassment, in contrast, is very much about such consequences[47] – about how one's standing and image will suffer once esteem has been diminished due to the embarrassing act. Other differences that have been discussed in the literature include the fact that shame generates a greater degree of emotional pain than embarrassment, that shame damages the core of one's self-understanding and self-worth far more than embarrassment does, and that feelings of shame linger longer than those of

[44] See Julien Deonna, Raffaele Rodogno and Fabrice Teroni, *In Defense of Shame: The Faces of an Emotion* (Oxford, UK: Oxford University Press, 2013).

[45] Braithwaite as discussed in Lu, Guilt and Reconciliation, p. 371.

[46] See, for instance, Jessica Tracy and Richard Robins, "Putting the Self into Self-Conscious Emotions: A Theoretical Model," *Psychological Inquiry* 15(2) (2004): 103–25.

[47] Fussi Alessandra, "Aristotle on Shame," *Ancient Philosophy* 35(1) (2015): 113–35, 8.

embarrassment.[48] While psychologists do not agree on the criteria for distinguishing between shame and embarrassment,[49] the common sense usage of the terms suggests that both emotions are largely social (namely they depend on an audience for their full appearance), that the identity of the audience (how many people, how much they matter to the subject) influences the intensity to which the emotion is felt, and that shame involves a greater degree of psychological damage to the agent than embarrassment. Furthermore, shame involves feelings of guilt for the violation of a moral norm and for the culpable infliction of harm, whereas embarrassment merely involves regret and annoyance at being exposed in a different light than one would have liked.

To the extent that Margalit is right, and humiliation *is* just refused recognition – the refusal to see someone as fully human, the denial of one's bodily autonomy, the dismissal of one's way of life – the disposition is not necessarily tied up with the feeling that something is wrong with one's character, or with a sense of shame more broadly. On the recognition-based understanding of humiliation, it is perfectly coherent to be humiliated and not feel anything is wrong with oneself or with the group that one belongs to. In fact, this is exactly why feelings of humiliation sometimes lead to violent recriminations – when people feel that they are being perceived, evaluated, and described unfairly they often push back forcefully. If feelings of shame and guilt hinge on a sense that one has violated a norm, fallen short, or hurt someone – humiliation does not. Conceptually, it is possible to feel humiliated without feeling guilty or ashamed. A group of soldiers that were captured after putting up a valiant fight may be mistreated by their captors – denied the means for basic hygiene, and forced to eat from the floor. This treatment is humiliating and, if the soldiers are not unusually stoical, they will feel humiliated, but there's nothing in the circumstances to suggest that they have to feel ashamed or guilty. Similarly, members of a severely mistreated minority (blacks under apartheid or Jim Crow, for example) can be subject to humiliating treatment by the authorities (segregation of schools and public facilities, unequal services) but they have nothing to feel guilty or ashamed about and, in fact, the humiliating treatment can generate a sense of defiant pride and help cement a sense of political identity.

[48] For a useful overview, see Ray Crozier, "Differentiating Shame from Embarrassment," *Emotion Review* 6(3) (2014): 269–76.

[49] Ibid., p. 275.

Yet, members of humiliated groups often internalize the humiliating views that their oppressors hold of them. And when that happens, feelings of humiliation and shame fuse. During World War II, Stalin and many of his magnates urged treating surrendering Soviet soldiers as traitors, and some of those captured may well have accepted that definition. And we know that some political prisoners during Stalin's Terror, though devoted Bolsheviks who were imprisoned for no good reason, started feeling ashamed and guilty, and not only humiliated, after a few weeks in Lavrentiy Beria's dungeons.[50]

This fusing between humiliation and shame is at the heart of Fanon's account of subjugation in *The Wretched of the Earth*. Colonialism, he tells us, involves dehumanization – sometimes manifested in the pretense that the native population simply isn't there, or in viewing it as animalistic, stupid, less than human. That does not mean that these humiliating attitudes must morph into a sense of shame on the part of indigenous populations. There's no reason for them to take on the colonizer's view. But in practice, Fanon argues, they often do. Humiliating others tends to generate shame and self-loathing. Indigenous populations, he writes, often internalize racist, essentialist views held by the colonizers "because after all 'we are a bad, quick tempered, aggressive people . . . and that's the way we are'."[51] Furthermore, the humiliating living conditions that colonialism breeds, the struggle to eat and make it through the day ultimately turns the colonized against each other: "[T]he sole objective is the need to fill that ever-shrinking stomach . . . who do you take it out on? The French are down on the plain with the police, the army and their tanks . . . inevitably you stumble up against yourself. Here lies this core of self-hatred that characterizes racial conflict in segregated societies."[52] As far as Fanon was concerned, the struggle against colonialism – the violent struggle – rehumanizes the native population; it gives back a sense of agency and freedom and makes the indigenous into agents again – perhaps in the eyes of the colonizers (who will now see them as enemies to be reckoned with as the French saw the Algerians after the FLN launched its fight) but much more importantly in their own eyes. The violent fight for liberation elevates the native population from the struggle for survival imposed on it, and in doing so, allows people to stop seeing themselves as organisms fighting to

50 A moving account of these dynamics can be found in Simon Montefiore, *Stalin: The Court of the Red Tsar* (New York: Vintage, 2005).
51 Frantz Fanon, *The Wretched of the Earth* (New York: Grove Press, 2004): p. 228.
52 Ibid., p. 232.

obtain enough calories and start seeing themselves as human beings defending a just cause. That struggle is appealing, in part, because it reverses the feelings of humiliation and shame that attend them.

If we are to follow up on the suggestion, articulated earlier, that humiliation often consists in a sense of being replaced, we will again reach a similar conclusion: Humiliation does not have to come with a sense of shame, but it often does. Consider a person who feels replaced by their spouse's choice of a new partner or consider someone who buys into the rhetoric of political replacement promoted by ISIS or white supremacist propaganda. The humiliation here springs from a sense of being dislodged or moved from a spot of special status or ascendency. That removal is seen to be unfair and injurious. Or it can even seem morally neutral – the result of economic forces larger than ourselves – as perhaps those who are replaced by automation feel. What's humiliating is the sense of being replaced – and that sense can easily morph into shame if it amounts to a perception that one is not good enough to keep their place, not good enough to maintain their status.

The distinction between humiliation and embarrassment seems more straightforward on the face of things. If Margalit is right that humiliation is, at least in part, the result of misrecognition by others, embarrassment is the realization that one has not lived up to or behaved according to the manner in which they are usually recognized (or would like to be recognized). Embarrassment is often self-inflicted, unlike humiliation, and, in either case, it is a far milder affective response – typically felt less intensely and for a shorter duration than humiliation. Normatively speaking, one should not feel humiliated when a gap is pointed out between their self-understanding and their behavior. The person who thinks themselves honest and decent should be embarrassed rather than humiliated when they publicly fall short. But, as we've noted earlier, people are often humiliated when they shouldn't be. Moral sentiments can exceed their morally prescribed intensity. We can feel humiliated when we should only feel embarrassed, and feelings of embarrassment can bring feelings of humiliation in their train if the agent has a hard time letting the former go.

Further complicating the distinction between humiliation, shame, and embarrassment is the curious fact that there is a phenomenological difference between trying to bring about these dispositions and actually suffering them. While being humiliated, ashamed, and embarrassed differ in the ways just described, shaming, embarrassing, and humiliating someone else are closely related intentions. This is probably because the intentional acts are all focused on injuring another's sense of dignity:

If someone actually *wants* to make another feel ashamed, humiliated, or embarrassed, they are probably not concerned with nuances about the degree of injury these dispositions involve. That shaming, humiliating, and embarrassing are more closely related than being ashamed, humiliated, and embarrassed suggests that the feelings have independent psychological contents that go beyond the intentions involved in inducing them. This conclusion bolsters our earlier observation that there must be more to feeling humiliated than simply being denied recognition. Just like there is more to being ashamed and embarrassed than a reflection of the intentions of the person shaming or embarrassing you formulated.

4.5 Humiliation and Technology

The rise of social media adds an interesting twist to the history of humiliation. The easy dissemination of information on social media platforms and the remarkable staying power of that information make it exponentially easier to humiliate people, and it also makes the effects of humiliation harder to shake. The journalist Jon Ronson recounts the case of a violinist who, after surviving a train crash that killed other fellow passengers, demanded in a tweet that Amtrak, the company which operated the train, retrieve her violin. The reaction from Twitter users – swift, harsh, and relentless – guaranteed that the violinist's lapse of judgment reverberated around the world for a very long time. While the violinist's tweet was certainly tone-deaf, it is far from clear that it merited the long-lasting legacy of humiliation that it did in fact generate.[53]

The ubiquity of smartphones and the ease of posting materials to social media means that acts of wickedness, stupidity, or indiscretion are likelier to get captured and disseminated. To put it simply, the chances that any of us will be humiliated because of something they said or did are exponentially greater than they were in the beginning of the twenty-first century.

Further, the very fact of easy, widespread dissemination of humiliating information calls up the need for a response from injured parties. The social dynamics here are not unlike those involved in dueling, just on a much broader scale: A challenge in public requires a response if the challenged party is going to save face. Consider Peter Singer and

[53] Jon Ronson, "How the Online Hate Mob Set its Sights on Me," *The Guardian* (December 20, 2015).
www.theguardian.com/media/2015/dec/20/social-media-twitter-online-shame.

Emerson Brooking's description of mutual provocations on social media between rival gangs:

> The decentralized technology thus allows any individual to ignite this cycle of violence. Yet by throwing down the gauntlet in such a public way, online threats have to be backed up not just by the individual, but by the group as well. If someone is fronted and doesn't reply, it's not just the gang member but the gang as a whole that loses status. The outcome is that anyone can start a feud online, but everyone has a collective responsibility to make sure it gets consummated in the real world. One death can quickly beget another. Sometimes, the online memorialization of the victim inspires vows of revenge; other times, the killers use it to taunt and troll to further bloodshed. One teen explained, "Well, if you're a rival gang, I'll probably send you a picture of your dead boy's candles (at a public memorial) ... or I'll take a picture by your block with a gun and say, 'Where you at?' It depends on how you take it ... and things will go from there.[54]
>
> Social media not only makes it easier for us to humiliate or be humiliated, it also increases the chances that humiliations will escalate.

To these dynamics, we must add a comment on the economy of attention online: Since humiliating moments often go viral (given the tested virality of strong emotions[55]) and since people want to go viral for a variety of reasons (whether having to do with a need for affirmation or a desire for monetary gain or both), the very existence of social media creates its own incentive to humiliate. One is dismayed (but no longer surprised) to see, over and over, that the first response of many bystanders to a humiliating situation is to record it rather than alleviate the distress of someone who is suffering.

4.6 Summary

We began this chapter by discussing Avishai Margalit's misrecognition-based theory of political humiliation. For Margalit, humiliation is primarily understood as the culpable denial of self-respect. His theory of humiliation has a normative emphasis. There are, in other words, morally better and worse reasons for feeling humiliated, and the simple fact that one does feel humiliated does not mean that the sentiment is warranted. Margalit

[54] Peter Singer and Emerson Brooking, *LikeWar: The Weaponization of Social Media* (New York: Eamon Dolan, 2018): pp. 13–14.

[55] See, for example, Jonah Berger and Katherine Milkman, "What Makes Online Content Viral?" *Journal of Marketing Research* 49(2) (2012): 192–205.

notes that political humiliation usually takes one of three forms – removing people from the human community (as when we liken them to animals), the negation of control (as in torture), and ignoring or looking through others.

After providing an account of this theory, we argued that Margalit does not sufficiently consider the contagious nature of political humiliation nor the possibility that the feeling might be present even when recognition is offered or, adversely, that we might be humiliated even by those whose recognition we don't want. In other words, we argued that phenomenologically there are reasons to doubt the connection Margalit draws between humiliation and recognition. We also looked at the conceptual differences between humiliation, shame, and embarrassment. We noted that despite these clear differences the way these emotions are experienced sometimes feels similar. In other words, feelings of shame, embarrassment, and humiliation can often fuse. We concluded the chapter with a discussion of the role of technology and, in particular, social media on the character of contemporary political humiliations.

CHAPTER 5

Humiliation in Foreign Policy

Hitler's rise to power was fueled by resentment over Germany's perceived humiliation at the hands of the victorious World War I Allies. The terms of the Versailles Treaty were, many Germans thought, haughtily dictated to them; the blame for the war's outbreak and its costs squarely placed on their shoulders. More recently, both Russia's and China's revanchist foreign policies have been justified by claims of historical humiliation – a version of history in which these once great powers came to be dismissed and disregarded by the United States and other ascendant Western nations. In the last decade, ISIS fashioned itself as the force that would reverse centuries of humiliation of Muslims by "apostates." The recreation of the historical Caliphate, which ISIS touted as its greatest achievement, was supposed to end the humiliation of the *Umma*. Examples can be easily multiplied: French resentment over humiliation in the Franco-Prussian War animated retaliation against the Germans; Egyptian and Syrian humiliation in the 1967 Middle East War fueled the surprise combined Arab attack on Israel in 1973.

That humiliation in the international arena is readily deployed as a tool for political radicalization and recruitment suggests that states have a national security interest in understanding what constitutes a humiliating foreign policy, realizing the risks involved, avoiding humiliating policy decisions when possible, and redressing past humiliations when appropriate. Doing all of this will not make groups like ISIS recant their claims, it will not make leaders like Putin abandon their ambitions either, but it would challenge the credibility of such actors' assertions.

5.1 What is a Humiliating Foreign Policy?

So what is a humiliating foreign policy? Though Margalit's theory of humiliation, taken up in Chapter 4, centers on domestic cases, some of the features he discusses are relevant for answering the question. Foreign policies that dehumanize, look through, and negate the self-control of other groups (or of members qua members of that group) are humiliating. But that does not take us far enough. What does it mean to dehumanize, ignore, and negate self-control in foreign policy? What are the manifestations of these tendencies in the relationships between groups?

One prevalent expression of these propensities is **acting vis-a-vis others simply because one can count on their inability to respond**, throwing one's weight around arbitrarily, imposing one's will, knowing all the while that one can't be counteracted. Thucydides' famous Melian Dialogue provides the paradigm case: The Athenians decided to overrun the small island nation of Melos because they were worried about the perception of Melian neutrality or sympathy towards Sparta. That Melos did not actually harm the Athenians or challenge them was immaterial. What mattered was that Athens had an interest in subjugating the island, that it had the ability to do so, and that Melos could not do anything about it. Thucydides' famous rendering of the Athenian position is worth quoting: "For our part, we will not make a long speech no one would believe, full of fine moral arguments – that our empire is justified because we defeated the Persians, or that we are coming against you for an injustice you have done to us . . . Instead, let's work out what we can do on the basis of what both sides truly accept: we both know that decisions about justice are made in human discussions only when both sides are under equal compulsion; but when one side is stronger, it gets as much as it can, and the weak must accept that."[1]

To humiliate in international relations – to dehumanize and negate the self-control of other nations – is to do what we want to them simply because we can, because it suits our interests, because we're powerful enough to ignore the status of those on the receiving end of our decisions. To be humiliated in foreign policy is to be done to, to be disposed of. That is what it must have felt like to be Melian under the circumstances Thucydides describes. That is what it must have felt like living in the Algerian Sahara, when France decided to test its nuclear weapons there in 1960. General Charles Ailleret explained that "the total absence of all signs of life" was "essential in choosing the site."[2] And yet those living in the town of Reggane, not too far away from the blast site, suffered the consequences for decades to come.

[1] Thucydides, *On Justice, Power, and Human Nature: Selections from the History of the Peloponnesian War*, trans. Woodruff (Indianapolis, IN: Hackett, 1993): p. 103.

[2] www.bbc.com/news/world-africa-56799670.

Shock and awe campaigns provide more recent illustrations of these dynamics of humiliation. In such campaigns, one actor inflicts immediate, massive, disproportionate damage on its enemies in order to sap their will to fight. But that message is sent at the expense of many uninvolved individuals who simply feel a massive force being applied to them as a strategy of political communication with someone who is not them. This was true in the shock and awe overture that opened the 2003 Iraq War,[3] and it was true in Israel's disproportionate response to the 2006 attack on it by Hezbollah – a response summarized by then IDF Chief of Staff Dan Halutz's promise to "turn the clock 20 years back on Lebanon."[4]

These naked displays of force are at the heart of humiliation in international relations,[5] and they create the kind of rage that often comes back and haunts the powerful. Indeed, the Melians warned the Athenians of just that. The fortunes of war are changeable, they told the Athenians, and simply applying power to another party because one can is dangerous, as there are no guarantees that political advantage will last: "[I]f you ever stumble you might receive a terrible punishment and be an example to others."[6] The Athenian generals took no notice.

Another feature typical of a humiliating foreign policy, another way in which we dehumanize, ignore, and negate self-control in international relations, is to **disregard the national heritage or history of others**. This was on display when the Americans allowed the Iraqi National Museum and National Library to be looted, while making sure to secure the country's oil ministry during the early days of the 2003 occupation. It has been on display in repeated pronouncements, over the years, by Israeli officials who, in response to demands to create a Palestinian State in the West Bank and Gaza, claimed that there was no such thing as a Palestinian people.[7] It was on display in the Saudi bombing of the Dhamar Regional Museum in Yemen – an attack that destroyed many of the country's priceless ancient artifacts.[8] And, it was, of course, on

3 www.cbsnews.com/news/iraq-faces-massive-us-missile-barrage/.

4 http://news.bbc.co.uk/2/hi/middle_east/5173078.stm.

5 For a discussion of the moral and practical risks involved in a strategy of intentional disproportionality, see Nir Eisikovits, "Proportionality and Self-Interest," *Human Rights Review* 11 (2) (2010): 157–70.

6 Thucydides, *History*, p. 103.

7 The most famous version of this claim was put forward by Israeli Prime Minister Golda Meir: "It was not as if there was a Palestinian people in Palestine and we came and threw them out and took their country away from them. They did not exist." www.haaretz.com/.premium-17-golda-meir-quotes-on-her-117th-birthday-1.5356683.

8 www.washingtonpost.com/news/worldviews/wp/2015/06/05/the-world-may-be-ignoring-the-destruction-of-cultural-treasures-in-yemen/.

display in the Taliban's demolition of the famous Buddhas of Bamiyan in Afghanistan and in ISIS' decimation of the ancient Syrian site of Palmyra.[9] Since national heritage and shared history are constitutive of group identity, such conduct is not only an affront to the significance of that identity but also, in the case of the physical destruction of key artifacts, undermines the conditions for its creation and maintenance. Of course, this kind of behavior is a special case of the arbitrary use of force described above. By dismissing or destroying a victim's national heritage, an aggressor is, in effect, communicating the following: "I am going to ignore or destroy the very heart of your history – either because there is no reason for me to care sufficiently (you are not important enough for me to avoid doing so) or because I know I can get away with it." Again, it is the feeling of helplessness in the face of the callous use of force that anchors humiliation in such cases.[10]

A third and related feature of a humiliating foreign policy is the tendency to **dismiss and ignore the complex social and cultural realities making up the lives of those we come into contact with**. The colonial powers that divided the areas of the "Levant" after World War I were ignorant of the intricacies of ethnic and tribal tensions in the area, and sliced up the area on their maps as if these realities did not exist or did not matter. There is, perhaps, no more powerful illustration of this tendency than the remark of Britain's Lord Sykes while negotiating with the French over control of the area that makes up contemporary Iraq and Syria: "I should like to draw a line," he told his French counterpart, showing him a map of the region, "from the 'A' in Acre to the 'K' in Kirkuk."[11] Part of that line became the border between Iraq and Syria. That the line made no

[9] www.mei.edu/publications/death-buddhas-bamiyan and www.smithsonianmag.com/smart-news/ancient-city-palmyra-gravely-damaged-isis-may-reopen-next-year-1-180970160/.

[10] An interesting question in this context has to do with the role of relative power in such instances of cultural dismissal: What if the party being dismissive is much weaker than the party it dismissed? Consider an example: Just as some Israelis have been claiming for decades that there's no such thing as a Palestinian people, some Palestinians have been claiming that the Jews have no historical relationship to the Holy Land. If the first claim is humiliating, isn't the second? This raises the question of whether a power differential is constitutive to political humiliation between groups. Prima facie, in the second case, the stronger party's sense of agency should not be hampered by the offensive statement nor should it feel deeply threatened by it – the pronouncement can be insulting, rude, even anti-Semitic. But in the absence of any power to act on it, it is questionable whether it can be humiliating. But the dynamics of asymmetrical power are more complicated than that. The weak often maintain the ability to elicit significant support from other, stronger supporters. Their refusals of recognition may have real consequences when, for example, they (the weak) are joined by other nations who affirm their position. We need not settle this question here.

[11] See James Barr, *A Line in the Sand: Britain, France and the Struggle for the Mastery of the Middle East* (New York: Simon and Schuster, 2011): p. 12.

ethnic and tribal sense did not matter. And, as anyone who follows the news from the Middle East can easily confirm, that straight line has haunted the lives of Shiites, Sunnis, Kurds, Yazidis, Druze, Alawites, and members of other religious and ethnic groups in the area ever since. The so-called "percentages agreement" of 1944 in which Stalin and Churchill carved up areas of influence in Eastern Europe over drinks displays the same disregard for those whose lives would be impacted. Churchill's later designation of the agreement as a "naughty document" in his memoirs, rather understates its impact and its coldness.[12]

The Americans arrived in Vietnam knowing little about the history of economic and political animosities between the Chinese and the Vietnamese in that country. They arrived in Afghanistan and knew virtually nothing about the tensions between the Peshmerga and Uzbeks.[13] In both cases, they devised military strategy and policy as if these nuances and local factors didn't exist or didn't matter. The message was clear: All the "locals" were the same; one didn't need to know what matters to them in order to control them.[14] Frantz Fanon has argued that such "systemized negation" of the locals was constitutive of colonial attitudes. Referring to the French colonization of Algeria, he wrote: "In Algeria there is not simply domination but the decision, literally, to occupy nothing but territory. The Algerians, the women dressed in haiks, the palm groves and the camels form a landscape, the natural backdrop to the French presence."[15]

To these three manifestations of humiliation in foreign policy, we might add two more: **failing to take another group seriously as a subject of**

12 See Mark Gilbert, "From War to Cold War," in Klaus Larres (ed.), *A Companion to Europe since 1945* (Hoboken, NJ: Blackwell, 2009).

13 Amy Chua, "Tribal World: Group Identity is All," *Foreign Affairs* (July/August 2018). Chua relays a striking exchange which illustrates this cluelessness about local attitudes and commitments. In 1997, Robert McNamara who served as secretary of defense during the Vietnam War met Nguyen Co Thach, who was the Vietnamese foreign minister. Thach was puzzled at the ignorance that seemed to characterize the American assumptions fueling the war. "You must never have read a history book. If you'd had, you'd know we weren't pawns of the Chinese," he told McNamara. "Don't you understand that we have been fighting the Chinese for 1,000 years? We were fighting for our independence. And we would fight to the last man … And no amount of bombing, no amount of U.S. pressure would ever have stopped us."

14 See Rory Stewart and Gerald Knaus, *Can Intervention Work?* (New York: Norton, 2012). In the book, the authors argue that this imperious approach now infects even those who come to foreign policy with good intentions; the authors claim it has been effectively standardized by international aid and development agencies and organizations, who believe that following the best practices and procedures of a generalized discipline of international development outweighs the importance of detailed knowledge of the language, customs, commitments, and history of the locals.

15 Fanon, *The Wretched of the Earth*, p. 182.

promises and agreements and **viewing another political entity primarily as a repository of resources** – to be pilfered for what it has to offer rather than to be considered as a group of people with a meaningful relationship to their country and its treasures. The English and French dismissal of their promises to the Arab tribes who helped them defeat the Turks during World War I is an example of the former. So is the long history of the American federal government's disregard of treaties with native nations. The race to control oil fields of the Middle East after World War I, Britain's desperate clinging to its "Crown Jewel" in India, and the Soviet exploitation of Ukraine's grain production capacities are examples of the latter.

It should be obvious that these five features of humiliation are not neatly distinguishable. When we don't take another's heritage seriously, when we care about them primarily for their resources, when we don't bother to study local complexities, it becomes far easier, and even natural, to exercise power arbitrarily. When we do exercise power over others arbitrarily, we often erase part of their heritage and cultural complexity such that there's less to notice even if we were inclined to pay attention.

Two final observations are in order about humiliation in international affairs. First, even if one agrees that it is both wrong and counterproductive, a humiliating foreign policy may be difficult to reverse. It is, for example, probably more difficult for an American administration to reverse a humiliating foreign policy than a humiliating domestic one that, say, disproportionately targets Blacks and Latinos for frisk and search policing. There is a political risk for elected officials in maintaining humiliating domestic policies. Alienating large swaths of voters by upholding such policies can eventually cost votes. Humiliating foreigners doesn't lose elections, or at least it doesn't right away. And, of course, in addressing a humiliating foreign policy there's the political risk of seeming weak or overly apologetic (or, making oneself vulnerable to being portrayed as such by one's political opposition). Barack Obama received tremendous pushback when, in a series of public addresses (most famously his 2009 Cairo speech), he admitted that the United States had not always been a force for good in the world. He was described as apologizing for America, an accusation that would figure prominently in later election cycles.[16]

[16] For an early repudiation of Obama's "apologetic" tone, see http://s3.amazonaws.com/thf_media/2009/pdf/wm_2466.pdf. For an overview of this theme as the 2016 elections approached, see www.washingtonpost.com/news/the-fix/wp/2016/05/10/obamas-trip-to-hiroshima-and-the-looming-apology-tour-narrative/?utm_term=.dd8d2b9cc71e.

In other words, while there are meaningful political incentives to reverse domestic humiliations (for example, losing a generation of Latinos as potential voters for one's party), there are no similarly powerful inducements to address international humiliations. It is easier (and clearly correct) to blame ISIS for attacks than to assess the role of a history of humiliation in the organization's recruitment successes.

Second, the distinction between humiliating foreign and domestic policies is not always clear. Decrees such as the Trump administration's Muslim travel ban or the Israeli decision to cancel the official status of Arabic as part of the state's recent Nationality Law are humiliating domestic policies. And yet these can be readily exploited by foreign actors for recruitment and other propaganda purposes. ISIS has proven it can easily use the domestic humiliations of Muslims in various countries as fodder for recruiting fighters or for facilitating the radicalization of lone-wolf terrorists.[17]

5.2 The Phenomenology of Political Humiliation

To be humiliated, as we've just noted, is to feel done to, looked through, to have one's culture and historical legacy ignored, to realize that one is not taken as a relevant subject of promises or concern, to feel that one's land is of interest primarily for the resources that can be extracted from it. Is there anything else to this disposition?

As we noted in Chapter 4, a key feature of the phenomenology of humiliation – in international relations, domestic politics, and even intimate relationships – has to do with anxieties and frustrations about replacement. I'd like to make a more detailed case for that statement but before we do so, a conceptual clarification is in order: I'm not claiming that the sense of being replaced is a sufficient or even necessary condition for the feeling of humiliation to set in. It is certainly possible to be politically humiliated without feeling personally or collectively displaced. The Germans felt humiliated at the 1919 Paris Conference not because the Americans, British, and French somehow "replaced" them on the world stage, but because they thought their interests were not taken seriously and they believed a disproportionate degree of responsibility for the Great War's harms was imposed on them. I am, however, arguing that in

[17] See, for example: www.washingtonpost.com/world/national-security/jihadist-groups-hail-trumps-travel-ban-as-a-victory/2017/01/29/50908986-e66d-11e6-b82f-687d6e6a3e7c_story.html?utm_term=.b43bb6ba6ba0.

political practice the sense of being replaced is very often central to the feeling of being humiliated.[18]

ISIS propaganda is replete with pronouncements about the lost glory of the *Umma* or Muslim nation – about honor lost to crusaders and other "infidels." Muslims, so the story goes, used to be warlike and widely revered, but have surrendered their ferocity and pride to other powers. Muslims used to be ascendant in war and culture but now they have become subservient to the West. As we shall see in Chapter 8, ISIS promised to change these trends.

The sense of replacement is also key to understanding Russia's recent foreign policy. From the annexation of Crimea to its involvement in Syria to the recent war in Ukraine, many of the country's actions have been framed as responses to being marginalized and dismissed after the collapse of the Soviet Union. Indeed, Putin and other Kremlin officials consistently claim that Russia has been relegated to second-rate power status after 1991 and that they (and only they) can halt this decline. Some of the primary complaints of the Russians include the NATO intervention in Bosnia in 1995 against their wishes, the quick dismissal of Russia's offer to help prosecute the "War on Terror" after the attacks of September 11 (specifically the offer to help with intelligence in Afghanistan), the beginning of the 2003 war in Iraq (again against Russia's wishes), and the second round of NATO expansion in 2004 which removed former Soviet Republics Estonia, Latvia, and Lithuania from Russia's traditional sphere of influence. We shall have to say much more about this example in Chapter 8.

Chinese foreign policy is similarly responsive to a long history of perceived humiliations. China's so-called "century of humiliation" refers to a period spanning the mid nineteenth to the mid twentieth centuries during which the country suffered a series of defeats in military engagements and was subjected to numerous disadvantageous treaties at the hands of Britain, France, Russia, and Japan. The CCP has used the "century of humiliation" to offer a narrative of Chinese decline – from the time of the "Middle Kingdom" (a name coined to denote China's pivotal and superior role in international affairs under the Qing Dynasty) to a marginal nation dismissed by neighbors and Western powers alike.

[18] The discussion here draws on Nir Eisikovits, "Political Humiliation and the Sense of Replacement," in Graham Parsons and Mark Wilson (eds.), *How to End a War* (New York: Cambridge University Press, 2023). For a discussion of whether this sense of replacement also operates in the context of workplace automation, see Jake Burley and Nir Eisikovits, "Workplace Automation and Political Replacement: A Valid Analogy?" *AI and Ethics* 3(4) (2022): 1361–70.

And, naturally, the role of the CCP – like that of ISIS and Putin in their own regional contexts – would be to restore China to its past glory.[19] In the aftermath of the deadly suppression of protests in Tiananmen Square, the CCP began the "Patriotic Education Campaign." The campaign aimed to educate China's youth about the country's humiliating experience at the hands of Western and Japanese aggressors and domestic "traitors," and to show them how "the CCP-led revolution had changed China's fate and won national independence."[20]

According to the most updated version of this narrative, China is now on the verge of reclaiming its rightful position in the world. Owing to Xi Jinping's emphasis on restoring national greatness, party leaders are portrayed as heirs to the dynasties of the eighteenth century, when Qing emperors demanded and received respect and even supplication from Western envoys.[21] This push makes sense of the formation of the Asian Infrastructure Investment Bank which allows China to take its fiscal future (and that of some of its neighbors) into its hands rather than depend on the Asian Development Bank, the World Bank, or other monetary institutions which are shaped by the United States. It similarly underpins the Belt and Road initiative, also known as "One Belt One Road" (OBOR), an ambitious infrastructure project impacting much of the region and about a third of the world's GDP.[22]

Finally, white supremacists in the United States, as well as their nativist, anti-migration counterparts in Europe, speak incessantly of the fear and humiliation of being displaced and replaced. "Jews will not Replace Us," the neo-Nazis famously chanted in Charlottesville in 2017. That chant found its inspiration in a book by a far-right French writer Renaud Camus. The influence of the latter's *The Great Replacement* has been wide-ranging. [23] As James McAuley recently wrote in *The Nation*:

[19] See Anderson, *China Restored*; Zicheng, *China's Grand Strategy*; Zheng, *National Humiliation*. See also Shivshankar Menon, "How China Bucked Western Expectations and What it Means for World Order," *Brookings Order From Chaos Blog* (2016). www.brookings.edu/blog/order-from-chaos/2016/03/10/how-china-bucked-western-expectations-and-what-it-means-for-world-order/.

[20] Zheng, *National Humiliation*, p. 116.

[21] See Frevert, *Politics of Humiliation*, p. 160.

[22] Christina Müller-Markus, "One Belt, One Road: The Chinese Dream and its Impact on Europe," *Notes Internacionals CIDOB* 148 (2016); Hong Yu, "Motivation behind China's 'One Belt, One Road' Initiatives and Establishment of the Asian Infrastructure Investment Bank," *Journal of Contemporary China* 26(105) (2017): 353–68.

[23] Renaud Camus, *Le Grande Remplacement* (Paris, France: Chez l'auteur, 2011). The English version is titled *You Will Not Replace Us* and was published in 2018 by the same publisher. For background on Camus, see Simon Murdoch and Joe Mulhall, "From Banners to Bullets: The International Identitarian Movement" (London, UK: HOPE not hate Charitable Trust, 2019). www.hopenothate.org.uk/wp-content/uploads/2019/12/Identitarianism-Report.pdf. See also Cecile

> [T]he term 'great replacement' has become a rallying cry of white supremacists around the world—the demonstrators who stormed through Charlottesville, Virginia, in August 2017; the man who killed 11 worshippers at the Tree of Life synagogue in Pittsburgh in October 2018; and especially Brenton Tarrant, who perpetrated the New Zealand mosque attacks in March of 2019. Tarrant posted his own 'Great Replacement' manifesto online—before murdering 51 people.[24]

Camus claims that a country's national identity is a function of the shared cultural norms and heritage embraced by its population. Based on the (unsupported) assumption that such coherence and uniformity really exist or ever existed, Camus postulates that if another population is introduced into such a coherent context under circumstances that overwhelm the native population's growth rate, the identity or essence of the existing civilization will be replaced by that of the incoming one. The argument's second premise (the first has to do with the existence of "cultural essences") is that people – in their individual or collective capacities – resent, indeed hate, feeling replaced. As Camus put it in a 2017 interview with *Vox's* Sarah Wildman, "[T]he refusal to be replaced is a very strong feeling in man. It doesn't really need to be put into hearts and into minds."[25]

The sense of replacement has elicited wildly different responses from the actors we have surveyed. ISIS responds with murderous brutality against anyone who does not meet its standard of Sunni purity. Russia has expanded its footprint in the Middle East and lashed out murderously against Ukraine. China has launched ambitious infrastructure programs and has pushed, aggressively, for technological and military parity with the United States. White supremacists use the sense of being replaced to justify their xenophobia and racism and to throw their support behind nativist politicians.

But though responses to the sense of replacement vary, there are common themes that characterize how it operates: In all cases, there is **a feeling of lost status**, a perception of lost ascendancy, of being removed from importance and consequence – usually **by someone (or in favor of someone) not viewed as a worthy competitor**. China was once the "Middle Kingdom" in the heart of world affairs, but the "century of

Leconte, "The Socio-Political Career of the Expression "The Great Replacement" among Right-Wing Party Networks in Germany: The Case of the Alternative for Germany (AfD) Party," Politix 126(2) (2019): 111–34.

24 www.thenation.com/article/archive/renaud-camus-great-replacement-brenton-tarrant/.

25 www.vox.com/world/2017/8/15/16141456/renaud-camus-the-great-replacement-you-will-not-replace-us-charlottesville-white.

humiliation" pushed it aside. Russia was once a superpower, first in the race to space, a formidable competitor with the United States for nuclear supremacy, and yet, since 1991, it has been relegated to the minor leagues of international relations. White men in the United States and Europe once enjoyed unquestioned ascendency, but now they see people of different races and ethnicities gain influence and prominence. The Islamic Caliphate used to reign supreme across much of the world and Muslims used to lead in both military and cultural pursuits while Europe was a backwater. Now the power relation has been reversed.

Relatedly, in each case, there is **a sense of entitlement denied, of a promise – explicit or implicit in political culture – broken**, an identity-defining, long-counted-on birthright withdrawn. For the Russians and Chinese, it was the promise that they would always be the citizens of great nations – a greatness that would compensate and make sense of the sacrifices they were often called to make for the state. For white supremacists in the United States and elsewhere, it was that white men would always hold sway, that they would have the most influence in how their countries would be ruled.

The sense of replacement, then, is about the diminution of status, the loss of an entitlement to someone perceived as unworthy of it, and, importantly, it is **also about the desire to take the lost status back**. Surrendering to the sense of replacement is not simply sinking into a politics of nostalgia. It is, rather, taking on an *actionable* nostalgia involving the desire to recreate the specialness, to bring back the ascendency, to reclaim the stolen birthright. In other words, the story of political humiliation, insofar as it has to do with replacement, is often a story of being removed from a pedestal, losing the assumed, natural, taken-for-granted preeminence that was constitutive of one's very political identity. It is the story of watching someone else instilled in your place – someone who looks and speaks differently, someone not from here, someone whom you perhaps were accustomed to historically look down upon (as the Soviets dismissed the shallow and "materialist" Americans, and as the Chinese dismissed the "barbarian" Japanese and Westerners). It is also a story about the angry determination to win one's place back and return things to their "natural" order.

Finally, the feeling of being replaced – on the world stage or as part of the elite in one's country – gives one a sense that they are replaceable, disposable, worthless – it gives the lie to the rhetoric of political specialness, manifest destiny, and exceptionalism that so many of the political players we have discussed like to instill. And it's this recognition of being replaceable, inter-changeable that is humiliating – perhaps even more than the actual loss of entitlements and exclusive access to social and material goods.

Again, there is much more to political humiliation than the sense of being replaced. As we just mentioned, among other features, political humiliation consists in the feeling of being done to at will, of not being a worthy subject of promises, of mattering for purely instrumental reasons. Why, then, linger on the sense of replacement? I think there are at least four reasons:

First, the sense of replacement crosses the divide between the personal and the political. The selfsame feeling of replacement that drives interpersonal humiliation is present in politics. The betrayed spouse jettisoned in favor of another feels (among other things) that they have been replaced and discarded, that what they thought was a unique status in the regard of their partner was not unique after all. That they are disposable rather than "the one." And that sense of replacement is the source of a great deal of anger. The same dynamics apply with varying degrees of intensity in nonromantic friendships, when, for example, an old friend has been discarded or neglected in favor of a new one.[26]

The similarity with the sense of replacement reported in political settings is striking. What are we to make of this resemblance? What does it mean that the phenomenology of replacement straddles interpersonal and political contexts? For one thing, it may suggest something about the intensity and primacy of the sense of replacement: That this feeling travels pretty much unchanged between intimate and political settings might attest to its unique force. There are other psychological dispositions that are salient in both interpersonal and political settings. But in many of these cases, the transition from the personal to the political takes on a metaphorical aspect: Individuals can heal from an injury, and we also speak of "national healing," but what, exactly, does it mean for a nation to heal from an episode of mass atrocity or historical injustice? The term grows a great deal hazier as it travels to politics. Similarly, individuals can forgive offenses perpetrated against them, or reconcile with those who hurt them,

[26] On replacement in interpersonal relationships, see John Tooby and Leda Cosmides, "Friendship and the Banker's Paradox: Other Pathways to the Evolution of Adaptations for Altruism," in Walter Runciman, John Maynard Smith and Robin Dunbar (eds.), *Evolution of Social Behavior Patterns in Primates and Man,* Proceedings of The British Academy 88 (New York: Oxford University Press, 1996): pp. 119–43, and Jaimie Krems, Keelah Williams, C. Athena Aktipis and Douglas T. Kenrick, "Friendship Jealousy: One Tool for Maintaining Friendships in the Face of Third-Party Threats?" *Journal of Personality and Social Psychology* 120(4) (2020): 977–1012.

and, of course, we often speak about political forgiveness and reconciliation (especially after Archbishop Tutu framed and justified the work of the South African Truth and Reconciliation Commission [TRC] as an instrument to promote national forgiveness and reconciliation), but what does it mean for one group to forgive another? Reconcile with another? Do groups really heal, forgive, and reconcile in the same way individuals do? I am not arguing that it is impossible for polities to heal, forgive, or reconcile, and these dispositions are, of course, the subject of a rich psychological and philosophical literature.[27] I am, rather, making the more modest claim that in the transition from the psychological to the political these terms lose some of their clarity, that their application becomes fuzzier in a way that does not hold when it comes to the sense of replacement.

The phenomenology of the personal and political versions of the sense of replacement and the resulting anger, disappointment, and resentment appear remarkably similar. In addition to pointing to the intensity of the sentiment itself, these similarities suggest that research on interpersonal feelings of replacement may be directly instructive for understanding the political manifestation of the feeling (and vice versa) – more so than in the case of other emotions. But there's an important caveat: Though the feelings may be almost identical, the grounds for the feelings are not. While the sense of replacement and the resulting humiliation between intimate partners is often well-founded, while the scorned partner, for example, often has a moral claim on not being replaced (the explicit or implicit commitment to loyalty, for example, in a committed monogamous relationship), the sense of replacement in politics is often both empirically unfounded and morally groundless. American white men, for example, are not really losing their ascendency – their sense of grievance is largely manufactured by political manipulation.[28] And even if they were losing their ascendency, in political society, unlike in a marriage, citizens are not and should not be promised a unique preferential status.

[27] See, for example: Charles Griswold, *Forgiveness: A Philosophical Exploration* (New York: Cambridge University Press, 2007); Colleen Murphy, *A Moral Theory of Political Reconciliation* (New York: Cambridge University Press, 2010); Nir Eisikovits, *Sympathizing with the Enemy: Reconciliation, Transitional Justice, Negotiation* (Leiden, The Netherlands: Brill/Martinus Nijhoff, 2010); Nancy Sherman, *Afterwar: Healing the Moral Wounds of Our Soldiers* (New York: Oxford University Press, 2015).

[28] As of 2017, whites comprised 73 percent of the US population, and as of 2019, Christians comprised 65 percent. See "*Population, Population Change, and Estimated Components of Population Change: April 1, 2010 to July 1, 2020*," United States Census Bureau. www.census.gov/programs-surveys/popest/technical-documentation/research/evaluation-estimates.html.

Second, and closely related to this last observation, a focus on the sense of replacement helps us distinguish normative and descriptive aspects of political humiliation. Groups certainly have a normative claim against being humiliated – for example, when they are simply acted upon because they are powerless to resist, when they are dismissed as subjects of binding promises, or when their cultural heritage is ignored. The Arabs cheated out of British and French promises for self-determination were correct to feel humiliated. The same is true of Palestinians who were informed by Golda Meir that "there [is] no such thing as Palestinians,"[29] or of Iraqis who realized that the American security forces were far more concerned about defending their oil ministry than the national museum. These are all instances of people who have suffered from a humiliating foreign policy another nation had undertaken, and they don't only *feel* humiliated, they are *right to feel* so. But Russia does not have a moral right to keep its status as a great power, China does not have a moral right to remain the "Middle Kingdom" in perpetuity, and white men do not have a right to lasting social prestige and influence in the United States and Europe. The sense of replacement in all of these cases is probably psychologically real, the resulting humiliation is often deeply felt, and it is potentially explosive, but it is not morally justified – it does not arise from nor does it generate a legitimate moral claim.

Stated differently, there is often a difference between the moral status and the psychological intensity of the sense of replacement in politics. That sense does not necessarily generate a moral claim because, in politics, we do not normally have any political entitlements to remain in a position of ascendency or to be held in unique regard. And yet feelings of replacement, even when they are not justified, are real, the humiliation they generate is real, and it can signal real trouble to come. In international politics as well as in domestic strife, the rise of a sense of replacement, regardless of its moral status, is predictive of instability and violence. Complaints about political replacement can function, then, as an early alert system, a seismograph, if you will, foretelling significant political turbulence.[30]

[29] Quoted in Baruch Kimerling, *The Palestinian People: A History* (Cambridge, MA: Harvard University Press, 2003): p. XXVI.

[30] It is interesting, but beyond our purview, to apply this predictive tool to another context all together – automation and the changing nature of work. According to many estimates, automation is going to replace and displace between a quarter and a third of the workforce in the next thirty years. In large chain stores, we are already seeing many employees working alongside the machines which will ultimately replace them. These trends may well bring about feelings of replacement on an unimaginable scale and, if the preceding analysis is applicable, a great deal of political unrest. For an argument to that effect, see Burley and Eisikovits, Workplace Automation.

Third, understanding the sense of replacement can help us distinguish between different varieties of conservatism. In particular, it can help us to tell apart moderate political conservatism from the fundamentalist, dangerous variety. In a recent study, Edmund Fawcett distinguishes between the "right" and the "hard right."[31] The former is a broadly Burkeian prudent conservatism skeptical of grand social schemes and more protective of lasting and meaningful features of the status quo than averse to change itself. The "hard right," in contrast, is reactionary and nostalgic – gripped with a profound anxiety about the loss of status. The project of the hard right is not to counter revolutionary ideas with suggestions for gradual reform, not to reinterpret a tradition in a way that makes it cohere with current developments, but rather to resist anything that challenges the status of certain elites. The hard right wants to retrench, and, if loss of status has occurred, to return to a mythically pure and innocent time that preceded the loss. For Fawcett, the history of conservatism has as much to do with the tension between these two strands as it does with the struggle of both against progressivism.

A focus on the sense of replacement is helpful in distinguishing between these two brands of conservatism and indicating when we are dealing with the nostalgic and reactionary variety. As the examples we have discussed suggest, the feeling of being replaced is usually accompanied by a strong political ambition to turn back the clock, to return to Eden, to go back to how things were before standing was lost and disappointment set in. The presence of the sense of replacement is thus indicative of the reactionary conservative impulse – the virulent form of conservatism that seeks not simply to ease the impact of social changes and make sure key traditions survive it, but to subvert any kind of alteration, to frame all change as a zero-sum game which robs people of key features of their identity. White supremacists, Russian and Chinese nationalists, ISIS leaders – in their various ways and to various degrees – do not want to make room in the new world order for their views – they want to revert back to the old world order in which their views and way of life were uncontested. Fawcett's distinction helps us to understand that conservatism itself has always been torn between a gradualist politics – willing to defend the need for elites and skeptical of the utility of abstract theories for political life, and a far more militant politics of humiliation which weaponizes and exacerbates the sense of loss and grievance that groups can feel as their power wanes.

[31] Edmund Fawcett, *Conservatism: The Fight for a Tradition* (Princeton, NJ: Princeton University Press, 2020).

The politics of humiliation practiced by this "hard right" is dangerous in part because of how activating, how psychologically powerful feelings of replacement can be.

Fourth, and finally, understanding the sense of replacement gives us useful insight into the so-called "Thucydides Trap," according to which the rise of a new power in international relations often leads to a clash with established forces.[32] By some estimates, up to 75 percent of competitions between emergent and entrenched political actors result in armed conflict.[33] The examination of the sense of replacement suggests that the anxiety stirred among the established power is not so much about the rise of the new actor or the new way of life it portends, as it is about the loss of status that comes with that emergence – the sense of decline, replacement, and, ultimately, humiliation that go with no longer "counting." This insight both reverses the emphasis of the Thucydides Trap – the decline of our vigor rather than the rise of the newly vigorous is what makes us jittery – and offers a way out of it. If the rising power is clever and generous enough to rise in a way that allows the declining actor to either maintain some of its status or at least save face, some of the risk embodied in Thucydides' insight may be averted.

5.3 Political Humiliation and Autocracy

Having reflected on the conceptual features and phenomenology of humiliation in international relations, let us proceed to examine a third important feature of how humiliation plays out between nations, namely the manner in which different types of regimes are susceptible to this sentiment and how they use it as a political tool.

In his excellent documentary *The Human Factor*, filmmaker Dror Moreh tells the story of the failure of Middle East peace talks in the late 1990s and early 2000s through the eyes of American officials. Among other events, the film describes the negotiations that almost led to a peace deal between Israel and Syria in 2000. Syrian President Hafez al-Assad, the film recounts, was humiliated when details of the negotiations were leaked

[32] As Thucydides puts it at the beginning of *History*: "The real cause, however, [for the breakout of war] I consider to be the one which was formally most kept out of sight. The growth of the power of Athens, and the alarm which this inspired in Sparta, made war inevitable." Thucydides, *History*, book 1, section 23.

[33] For an interesting application of the Thucydides Trap to the Chinese American context, see Graham Allison, *Destined for War: Can America and China Escape Thucydides's Trap?* (Boston, MA: Houghton Mifflin Harcourt, 2017).

and made it seem like he made concessions to the Israelis, only for Israeli Prime Minister Ehud Barak to go back on a promise to withdraw from the Golan Heights. The leak showed Assad as weak before his people – it made him appear as if he gave something without getting anything in return. Assad's reported humiliation in this case highlights the relationship between political humiliation and a country's form of government. Do democratic leaders get humiliated in the same way that autocrats do? Probably not. Autocrats' power rests on the correspondence between the image they try to project and the reality. They are thus especially sensitive to that image being debunked. A democratic leader's power rests, in large part, on their legitimacy – on the fact that they were elected and received a popular mandate. Stated differently, autocrats are more vulnerable to (and as a result are more scared of) humiliation than democrats. Yes, a democratic leader can be humiliated as well – shown up as less smart or strong than they are trying to look like. But that gap can be counteracted through the ballot box. An autocrat's image cannot. A George W. Bush can appear buffoonish on television, mispronounce words, or even choke on a pretzel.[34] If he gets reelected because the public deems he is doing well enough economically, or because they deem they have no better choice, or because they ultimately don't know which leader can best serve the state, those humiliations don't really matter. A Putin, in contrast, left waiting for Erdogan[35] or seen to be losing in Ukraine is in a different situation. A Stalin contradicted in public by a staffer is vulnerable in the same way and a Mohammed bin Salman challenged by domestic critics is too. None of them has the buffer of demonstrable public support to rely on. As a result, autocrats will typically be more anxious to avoid humiliation and will react to it in a more visceral fashion. Needless to say, critics of all three strongmen just mentioned learned that lesson the hard way.

Sophocles captures this relationship between power and vulnerability to humiliation perfectly. Simply being powerful, the tragedy of *Ajax* suggests, makes you prone to being humiliated in a way that we mere mortals are not. As rumors of Ajax's madness and humiliation by the gods spread (Ajax was made to believe that he was fighting his enemies while he was actually slaughtering domestic animals), people began to gloat.

The Chorus warns that Odysseus is spreading these rumors and that people are believing him:

[34] www.latimes.com/archives/la-xpm-2002-jan-14-mn-22490-story.html.
[35] www.theguardian.com/world/2022/jul/20/vladimir-putin-waiting-erdogan-turkey-russia.

> Such tales as these, whisperings and fabrications,
> Odysseus is supplying to every ear.
> And many believe him. For as he speaks of you,
> His words win credit, and each new hearer
> Even more than the teller relishes his chance
> To gloat over your distress.
> Strike at a great man and you will not miss;
> But if one should bend such slander at me,
> None would believe him. Envy talks
> after magnates of wealth and power[36]

Those with great power, Sophocles suggests, are ironically also greatly vulnerable to humiliation – the very magnitude of their strength makes them larger targets for potential ridicule. It is more difficult to miss once one strikes at them. Upon learning about Ajax's fate, Tecmessa, his war captive cries out: "Ajax! To be so great and suffer this!"[37] but what her despondent exclamation misses is that Ajax suffers exactly because he is so great. His very prominence makes him more susceptible to suffering in the way he does.

Autocrats are not only more vulnerable to humiliation – they are also, of course, more prone to use it as a tool of governance. Since they don't really work through institutions or coalition building, one of the most effective instruments they have for governing is the projection of raw power. And it's hard to think of a more visceral way to do that than humiliating others. Stalin was fond of humiliating his magnates in front of visiting dignitaries, going around the dinner table, introducing his henchman by role, and announcing that he would execute each boss who failed to do his job. He was also fond of forcing his key officials to drink themselves into a stupor in public. Saddam Hussein reportedly had his underlings driven to an undisclosed location with their eyes blindfolded, frightening them into thinking they were going to be executed, only to award each with a handsome cash prize. In both cases the message was clear. It's as if the dictator was saying: "[A]ll of you are dispensable and I can do with you as I please. You serve at my mercy."[38]

5.4 The Dual Role of Humiliation

As we close this chapter, it's worth noting the dual role of humiliation in group conflict. Humiliation can serve both as a motivator to fight and as a tactic for fighting.

[36] Sophocles, *Ajax*, lines 150–58. [37] Ibid., line 923.
[38] On Stalin, see Montefiore, *Stalin*. On Saddam Hussein, see Marianna Yarovskaya, *How to Become a Tyrant*, Netflix Documentary Series (Estuary Films, 2021): episode 2.

One often-cited historical example of humiliation as a spurn to war is, of course, Germany's humiliation in Versailles and its role in the subsequent rise of the Nazis. The sanctions imposed on Germany involved a dismantling of its fleet, the requirement to build shipping for the Allies, the loss of its colonial holdings (while retaining liability for them), and also the confiscation of private property in those holdings. The Allies also imposed a great strain on Germany's coal resources and manufacturing capacity – the very lifeblood of the country's economy. John Maynard Keynes argued in his classic discussion of these stipulations in *The Economic Consequences of the Peace* that the treaty was backward rather than forward-looking; that the Allies were too distracted and small-minded to make a real effort to integrate Germany back into Europe and that the result of Versailles would be another war that would make the Great War pale in comparison.[39] For several decades after World War II, many historians agreed with Keynes, citing the humiliation at Versailles as an important cause of Hitler's rise. More recently, revisionist scholars have challenged this view arguing that the demands made on the Germans were not unusually harsh or out of proportion with the wrongs the Germans had committed. Perhaps the best well-known among these is Margaret MacMillan's *Paris 1919: Six Months that Changed the World*.[40] Whether or not the Germans *should* have been humiliated by the treaty's stipulations, it seems quite obvious that they actually were;[41] for many of them, these sentiments made Hitler's promises of restoring German honor all the more alluring.

On June 21, 1940, a week after the Nazis marched into Paris, Hitler ordered the "Compiègne Wagon," the train car in which the Germans signed the 1918 armistice, to be removed from the French museum in which it was kept. The wagon was taken back to the exact spot where the defeated Germans signed the agreement that ended the Great War. Hitler then marched a French delegation into the train car and dictated his terms for accepting their surrender. With exacting symmetry, the German humiliation in World War I was reversed.[42]

[39] John Maynard Keynes, *The Economic Consequences of the Peace* (London, UK: Empire Books, 2012): chapter 4.

[40] Margaret MacMillan, *Paris 1919: Six Months that Changed the World* (New York: Random House, 2003).

[41] See, for example, David Terman, "Paranoid Leadership" in Charles Strozier, Daniel Offer and Oliger Abdyli (eds.), *The Leader* (New York: Springer, 2011) and Louise Slavicek, *The Treaty of Versailles. Milestones in Modern World History* (New York: Chelsea House Publications, 2010): pp. 41–43, 58.

[42] www.iwm.org.uk/history/what-you-need-to-know-about-the-battle-of-france.

Then there are some of the other examples we have already mentioned – the humiliation of Arab armies by Israel in 1967, which spurned the Egyptians to plan their surprise retaliatory attack in 1973, the perceived and real humiliations of Russia by the United States and other NATO powers which Putin uses to justify his war in Ukraine, the history of humiliation that ISIS invoked in justifying its recreation of the Muslim Caliphate. Cases can be multiplied – from the French seething over humiliation in the Franco-Prussian War to Milosevic weaponizing a mythical fourteenth-century defeat to fan the flames in the Balkans. Grounded or not, claims of historical humiliation have a long history of moving nations to war. A historical humiliation allows interested political actors or "humiliation entrepreneurs," to deploy a memorable term coined by psychologist Evelin Lindner, to use the immediate or distant past as a battle cry that galvanizes national sentiment.[43]

But if humiliation is a cause of war, if the disposition can be weaponized to send people into battle, it is also a way of war, a weapon of war, a method for waging it. Psychologically, that is hardly surprising. Humiliation, writes Lindner, is a "nuclear bomb of emotions" that generates an overwhelming need for retaliation.[44] Those who have been humiliated humiliate others in turn, once they can, because they are replaying the script that was imposed on them, because they are trying to regain a sense of agency that their own humiliation extinguished, because they are acting out the rage that their humiliation instilled in them. As the psychiatrist Adam Phillips put it some years ago: "We know that some people's psychic survival – everybody's psychic survival some of the time – depends on their capacity to humiliate others, to make others experience what they have suffered, as a way to reassure themselves that they are now the humiliators rather than the humiliated ... to transform the trauma of vulnerability into the triumph of omnipotent control."[45]

Or as the poet WH Auden put it more simply:

> I and the public know
> What all schoolchildren learn,
> Those to whom evil is done
> Do evil in return.[46]

[43] The term is taken from Evelin Lindner, *Making Enemies: Humiliation and International Conflict* (Westport, CT: Praeger, 2006).

[44] Ibid., p. 32.

[45] Adam Phillips, "The Magical Act of a Desperate Person," *London Review of Books* 35(5) (2013).

[46] Wystan Hugh Auden, "September 1, 1939," in *Selected Poems* (New York: Vintage Books, 1989): p. 86.

But beyond this psychoanalytic explanation in which humiliating an enemy is a massive reenactment of past trauma, there are more straightforward tactical reasons to humiliate in war.

While humiliating one's enemies does seed long-term hatred and resentment, though it often precipitates the next war, it also, given the very nature of this disposition, negates agency and breaks the spirit of the humiliated. And a passive, broken adversary is easier to fight and control. Humiliation, on this explanation, is a widespread tool of war because it is as tactically effective as it is morally awful. One need only look at the sunken faces of captured soldiers marching in their tattered uniforms and without shoelaces to grasp the point– the humiliated are not likely to lash back – at least not immediately. Humiliation, to quote Lindner once again, revolves around "putting down and holding down"; it is an "enforced lowering of [a] person or group by a process of subjugation that damages their dignity."[47] And this lowering makes that person or group that much easier to control.

Furthermore, given its visceral psychological salience, humiliation is also a powerful agent of deterrence. If we watch a conquering force humiliate an enemy, we might be more easily convinced to surrender without a fight in order to avoid humiliation ourselves. Thus, the conquering force is incentivized to humiliate not only to make a specific rival easier to control, but also to send a message to potential future rivals. Finally, there is the question of humiliating as a psychological outlet. Morally vacuous military leaders, concerned exclusively with success, will embrace ways for their troops to vent their rage and feelings of helplessness so that they maintain operational efficiency. Permitting them to humiliate enemy soldiers and the civilians who back them – from public beatings to campaigns of sexual violence – has, historically, been a method of choice. Consider Stalin's blood-curdling answer when confronted by Yugoslav Politburo member Milovan Djilas about the widespread rape and pillage Russian soldiers were committing as they advanced into Germany: "You have of course read Dostoevsky?" he asked Djilas, "Do you see what a complicated thing is man's soul? . . . Well then imagine a man who has fought from Stalingrad to Belgrade – over thousands of kilometers of his own devastated land, across the dead bodies of his comrades and dearest ones? How can such a man react normally? And what is so awful about his having fun with a woman after such horrors?"[48]

[47] Lindner, *Making Enemies*, p. XIV.
[48] Montefiore, *Stalin*, p. 479.

5.5 Summary

We began this chapter by outlining what constitutes a humiliating foreign policy. The features of such a policy include the arbitrary application of coercion to weaker parties, a disregard for the national heritage or history of those we encounter, dismissing the complex social and cultural realities making up the lives of other groups, failing to take other nations seriously as subjects of promises or binding agreements, and viewing another political entity primarily as a repository of resources. We then lingered on a key feature in the phenomenology of political humiliation, namely the sense of being replaced. This sense manifests in a perception of being removed from importance and consequence – usually by someone not viewed as a worthy competitor. A person or group experiencing the sense of replacement will complain of a perceived historical entitlement denied, a promise – explicit or implicit in political culture – broken. The feeling is not merely nostalgic. It almost always involves a desire to take the lost status back. Those who feel replaced most often want to bring back the ascendency, to reclaim the stolen birthright. Finally, we noted that the feeling of being replaced gives one a sense that one is replaceable, disposable, worthless. After describing the phenomenology of replacement, we pointed out that it is particularly important to understand this sentiment because it straddles personal and political psychology, and because a focus on the sense of replacement helps us distinguish normative and descriptive aspects of political humiliation. Furthermore, understanding the sense of replacement can help us distinguish between different varieties of conservatism and give us insight into the so-called "Thucydides Trap" according to which the rise of a new power in international relations often leads to a clash with established forces. After discussing the phenomenology of replacement, we observed the difference between democratic and autocratic rulers in their susceptibility to humiliation. We concluded the chapter with a discussion of the dual role of humiliation as both a driver and method of war.

CHAPTER 6

Humiliation and Political Theory

In Chapters 4 and 5, we looked into the meaning of political humiliation and its place in foreign policy. In this chapter, I want to show that concerns about humiliation and its consequences already pervade our thinking about politics, that the concept has a latent role in political theory, and that making that role explicit can help us clarify important theoretical and practical commitments. In particular, I claim that a focus on the need to avoid humiliation informs arguments central to the ethics of how we fight wars (*jus in bello*), the ethics of appeasement, and questions in transitional justice. My purpose here is not to take a position in the ongoing debates in these fields. It is, rather, to show that (somewhat hazy) views about humiliation already influence those debates. I want, in other words, to show how concerns about humiliation shape political philosophy, even if political philosophers have not yet fully acknowledged this.

6.1 The Role of Humiliation in Just War Theory

Our discussion of the role of humiliation in radicalization processes brings into focus an old, pragmatic rationale for moral restrictions on the conduct of war – the branch of just war theory referred to as *jus in bello* in professional jargon. That rationale can be traced back to Aristotle's commonsense insight that we fight for the sake of peace just as we work for the sake of leisure.[1] Fighting in a way that destroys the chances of a later peace is perverse and senseless – the martial correlate of being a workaholic. Part of the moral basis and practical import of just war theory in general, and of *jus in bello* requirements in particular, is to avoid actions that make future peace impossible. On this understanding, precepts like noncombatant immunity or proportionality are meant not only to spare the innocent,

[1] Aristotle, *The Politics,* trans. Saunders (New York: Penguin, 1981): book 14.

but also, by limiting the harm we do, to facilitate future coexistence with those who are now enemies.

Humiliating one's adversaries can create a mindset prone to extremism even decades after the humiliating act. Stated differently, fighting a war in a way that humiliates the enemy is wrong because humiliation is wrong in itself. It is also harmful in its consequences: Humiliating an enemy makes it harder to come to an agreement with the enemy once the war has subsided. Humiliation embeds residues of anger which can be readily agitated to the surface. On this understanding, shock and awe tactics that seek to consciously undermine *jus in bello* restrictions of proportionality or noncombatant immunity in favor of deterrence are problematic, not only because they knowingly target those who are not involved in fighting, but also because they terrify and humiliate an adversary into submission. Quite often, whatever these tactics gain in short-term deterrence they lose in the seething, long-term rage they create.

When Hezbollah abducted two Israeli soldiers and bombarded northern Israel in 2006, Israel's chief of staff threatened that unless the soldiers were returned immediately, the IDF would begin a massive assault on Lebanon that would "turn back the clock" on it by twenty years. [2] The Israelis proceeded to make good on that "shock and awe" threat in the war that ensued.[3] The calculation that such a disproportionate response would make ordinary Lebanese seethe at Hezbollah rather than at Israel was, of course, risky. Countries that fight against guerilla organizations like to label their operations as efforts to "dismantle the infrastructure of terrorism."[4] But visiting humiliation on one's enemies (for example through shock and awe tactics) actually creates and cements that infrastructure, insofar as these humiliations endow a young generation of fighters with hate and the willingness to fight. Strategies of calculated overreaction may

[2] http://news.bbc.co.uk/2/hi/middle_east/5173078.stm.

[3] For a discussion of these examples and a realist rationale for military proportionality, see Eisikovits, Proportionality.

[4] Israel often labels its operations in those terms and accuses the Palestinian Authority of failing to dismantle these infrastructures. Consider Likud politician Silvan Shalom's speech to the UN in 2003: "Failure to dismantle Hamas has brought our diplomatic efforts to a standstill. Rather than acting to fulfill its obligation, the Palestinian Authority has chosen the route of inaction, and complicity in terror. We cannot allow this to continue. We must bring back the hope that we can build a better future for our children. The infrastructure of terror must be dismantled so that we can put our peace efforts back on track. There is no other time. There is no other way." The full speech is available here: www.un.org/webcast/ga/58/statements/israeng030925.htm. Similarly, Indian officials regularly accuse Pakistan of failing to dismantle anti-Indian terrorist infrastructure: www.rediff.com/news/2006/mar/02pak.htm.

or may not achieve significant military deterrence. But whether or not they work as deterrents, they also make it harder to change future attitudes.

Perhaps the most famous version of this argument is delivered by Immanuel Kant as part of his doctrine of right in *The Metaphysics of Morals* as well as in his famous essay "Towards Perpetual Peace." Preliminary article 6 in that essay reads: "*No State Shall, during War, Permit Such Acts of Hostility Which Would Make Mutual Confidence in the Subsequent Peace Impossible.*"[5] Kant explains that "some confidence in the character of the enemy must remain even in the midst of war, as otherwise no peace could be concluded and the hostilities would degenerate into a war of extermination." He mentions assassins, spies, and the use of poison as examples of measures that undermine trust.

A cursory glance at the news makes Kant's examples seem quaint. And yet the basic idea is straightforward and as relevant as ever: Regardless of what one *can* do to one's enemies, one *should* only do in war what is consistent with the state of war eventually coming to an end. As we saw earlier, the same lesson was reiterated by the Melians when they warned the obstinate Athenians that their viciousness would come back to haunt them. Machiavelli delivers a variation on the theme when he insists that instilling hate is not compatible with a prince's political survival. It is better to be feared than loved (if one must choose) he tells us, but it is still crucial not to be hated: "[T]he prince, nevertheless, ought to make himself feared in such a mode that if he does not acquire love, he then avoids hatred; for being feared and not hated can go very well together."[6] A prince who does become hated due to excessive cruelty "ought to fear everything and everyone."[7] As some of the examples we have discussed suggest, humiliation creates the kind of distrust and long-lasting hatred that can fuel a "war of extermination" long after the original insult is suffered.

[5] Immanuel Kant, "Perpetual Peace," in Thom Brooks (ed.), *The Global Justice Reader* (New York: Wiley, 2008). The scholar Paul Saurette argues that Kant's view of jus in bellum can be read to both allow humiliations like those visited in Abu Gharib and to prohibit them. This reading is predicated on the comment that Kant makes according to which it is allowed "to extract supplies and contributions from a defeated enemy." Insofar as the humiliating interrogations at Abu Gharib in the wake of the invasion of Iraq were meant to extract intelligence from Iraqis, the policy, so the argument goes, may be justified on Kantian grounds. But this seems to me to be a misreading of Kant. The plausible reading is that one is allowed to "extract the supplies and contributions" from one's foe as long as the way one does so does not undermine the possibility of future trust and the chances of future peace. It's implausible to think that Kant assumed that spies and the use of poison would be detrimental to the chances of future relationships but that forcing your enemies to stand naked in front of American guards or forcing them to simulate sexual acts would pass that same test. See Paul Saurette, *The Kantian Imperative: Humiliation, Common Sense, Politics* (Toronto, CA: University of Toronto Press, 2005): pp. 238–41.

[6] Machiavelli, *The Prince*, p. 101. [7] Ibid., p. 113.

Consider the two key expectations at the heart of *jus in bello* restrictions – the requirements of noncombatant immunity and proportionality. Ignoring these guidelines amounts to waging a war of humiliation – targeting civilians in order to achieve military purposes is a classic example of using moral agents as "means merely." To refer to our earlier theoretical discussion of humiliation in foreign policy, when we target civilians we treat them not according to their deserts but, rather, on the basis of the power differential between us – we act on them rather than interact with them – because we can. We see them not as subjects of promises with rich cultural heritages but as resources at our disposal. Similarly, when we apply disproportionate force in order to deter an enemy we are again using the excess suffering of soldiers, and, often, of uninvolved civilians, in order to send a message to their leadership. Area bombings, hiding munitions in hospitals, the taking of civilian hostages are all humiliating tactics, and as such they make the chances of achieving peace in the future slimmer.

Understanding why humiliation is detrimental to the prospects of peace requires a closer look at its psychological impact. Humiliation harms one's capacity to trust that others will follow basic precepts of decency in future interactions. It destroys what Jean Amery called "trust in the world." As he put it in his essay on torture: "I don't know if the person who is beaten by the police loses human dignity. Yet I am certain that with the very first blow that descends on him he loses something we will perhaps temporarily call 'trust in the world.' Trust in the world includes all sorts of things: the irrational and logically unjustifiable belief in absolute causality, perhaps, or the likewise blind belief in the validity of the inductive inference. But more important as an element of trust in the world and in our context, what is solely relevant, is the certainty that by reason of written or unwritten social contracts the other person will spare me – more precisely stated that he will respect my physical and with it also my metaphysical being. At the first blow, however this trust in the world breaks down. And the other person, opposite whom I exist physically in the world and with whom I can exist only as long as he does not touch my skin surface as a border, forces his own corporeality on me with the first blow."[8]

Reflecting on the writings of Jean Amery, Phil Leask argues that the act of humiliation both robs its victim of the capacity to trust and traps them in fear. Once humiliated, a victim will have "the sense that the world is not as he previously understood it to be and that he is vulnerable to unexpected, arbitrary acts by someone with more power."[9] The order of things,

[8] Amery, Torture, p. 28.

[9] Phil Leask, "Losing Trust in the World: Humiliation and its Consequences," *Psychodynamic Practice* 19(2) (2013): 129–42, 132.

the set of beliefs about how human interaction is supposed to proceed, breaks down. The humiliated party is angry, afraid, and often wants revenge. Humiliation, as Leask puts it, "leads to a predictable set of emotions . . . in which rage and the desire for revenge, combined with a sense of impotence, tend to dominate."

Humiliation harms the chances for future peace, then, because the psychological dispositions it instills – loss of trust, rage, vindictiveness, a sense of helplessness are all antithetical to political reconciliation. Quite simply, peacemaking, the process of turning an enemy into a partner, since it is risky, is predicated on the capacities to hope and trust; it requires hoping that the future can be better than the past, and it often requires a leap of trust – taking the political risks involved in compromise (such as a first withdrawal from contested territory or a first drawdown of forces).

As we noted in Chapter 5, humiliating one's enemies can be both psychologically compelling and, at least in the short term, militarily effective. Hence, the temptation to use humiliation as a weapon of war has, historically, been considerable. *Jus in bello* limitations try to rein in that temptation because, beyond the inherent damage humiliation inflicts on one's psyche (and in large part because of that damage), humiliating an enemy makes the cycle of violence likelier to continue.

There is another, more tenuous variation on this warning against humiliating one's enemies – another anti-humiliation rationale underpinning *jus in bello* demands for proportionality and noncombatant immunity. This one is based on considerations of self-interest. On this view, one is to refrain from humiliating an enemy not because of the damage humiliation does to the prospects of peace, but because of the psychological harms it brings on the humiliators themselves. The worry is that the tendency to humiliate inculcated by waging unlimited war will travel. Those who humiliate enemies will end up humiliating their own. These psychological tendencies cannot be switched on and off at will.

The late dissident Russian journalist, Anna Politkovskaya, described, in chilling terms, what happened when Russian soldiers returned from a tour of duty in Chechnya, where they practiced indiscriminate rape and torture: "After this lawlessness, they leave for their homes, all over the country. Chechnya as a mode of thinking, feeling, and acting spreads everywhere like gangrenous cells and turns into a nationwide tragedy, infecting all strata of society."[10] In other words, the disdain and disregard for individual

[10] The quote is taken from Michael Specter, "Letter from Moscow: Why are Vladimir Putin's Enemies Dying?" *The New Yorker* (January 29, 2007).

life that facilitates the Russian military's iron grip on Chechnya becomes a chronic trait of those who are involved, an acquired attitude that follows the soldiers home, informing their behavior outside of uniform.

Some realists make a distinction between domestic and international norms, holding that while moral and constitutional standards are irrelevant and dangerous in the conduct of foreign affairs, they are commendable and necessary in the domestic arena. This view, most clearly articulated in the famous Cold War Policy Paper NSC-68, informed the covert operations of Eisenhower, Kennedy, Johnson, and Nixon. It justified the financing and orchestrating of foreign rebellions, assassinations, and a host of more modest American "involvements" abroad. But the duality turned out to be untenable. John Lewis Gaddis summarizes the point thus in his history of the Cold War: "The authors of NSC-68 had assumed that there could be separate standards of conduct in these two spheres: that American leaders could learn 'not to be good' in waging the Cold War while remaining 'good' within the framework of their own domestic democratic society. It had been hard enough to maintain that separation during the Eisenhower and Kennedy years: both presidents had been forced to admit that their 'denials' in the U-2 and Bay of Pigs incidents had not been 'plausible.' With the Vietnam War the line between what was allowed overseas and what was permitted at home disappeared altogether. The Johnson administration found it impossible to plan or prosecute the war without repeatedly concealing its intensions from the American people."[11] What we might call the Politkovskaya Principle can be summarized in simple terms: It is impossible to be a Machiavellian abroad while remaining a Kantian at home. Patterns of foreign behavior quickly become norms of domestic conduct.[12]

The psychiatrist Frantz Fanon gives us an account of the self-devouring dynamics of humiliation as they manifest in the colonial setting. The tendency to humiliate not only spreads to those around the humiliator, but also ultimately turns on the perpetrator himself. In a series of fascinating dispatches, Fanon documents the residue that torture leaves on its perpetrators (the French in Algeria, in this case), and the psychological destruction it leaves in its wake. An interrogator cannot stop hearing screams, and must sleep with cotton balls in his ears or with the radio on. Another police inspector who was involved in violent interrogations feels mounting aggression towards almost anyone he meets outside of

[11] John Lewis Gaddis, *The Cold War: A New History* (London, UK: Penguin Books, 2005): p. 171.

[12] I argue elsewhere that the Politkovskaya Principle constitutes an important argument for adhering to the *jus ad bellum* and *jus in bello* requirements of proportionality. See Eisikovits, Proportionality.

work: “He can’t put up with noise. At home he has a constant desire to give everyone a beating. And he violently assaults his children, even his twenty-month-old baby.”[13]

6.2 Humiliation and Transitional Justice

The term “transitional justice” denotes an interdisciplinary field of academic study as well as a set of institutional practices focused on how countries can best come to terms with their past – acknowledging and redressing past wrongs. Part of the promise of transitional justice measures is to reintroduce the significance of political recognition into societies emerging from a history of mass atrocity. What war crimes tribunals such as the Nuremberg trials do, in addition to holding people accountable for past crimes, is to make victims who used to think that they didn’t merit the law’s protection – a humiliating feeling (especially if we understand humiliation as rejected recognition) – realize that their status has changed. What truth commissions like South Africa’s TRC do, in addition to setting the historical record straight, is to provide victims a forum for publicly speaking about what has happened to them – a forum that is centered on them (more so than trials) and thus undoes some of the past humiliation of being voiceless. What lustration policies or administrative purges do (such as the denazification policy imposed on Nazi Germany) is to end the sense of impunity so typical of authoritarian regimes (and so humiliating to those who are subject to such regimes). Lustration communicates that those who practiced unbridled power can no longer do so. In communicating this, it works to undo the humiliation attendant to powerlessness before authority while also sowing the seeds of trust in the new government.[14]

[13] Fanon, *The Wretched of the Earth*, pp. 196–97. Fanon’s account and our general point here are reverberations of Socrates’ classical argument, reiterated in the *Gorgias*, the *Republic*, the *Crito*, and elsewhere – that injustice ultimately destroys the person committing it. While I don’t want to enter the discussion about what Socrates meant by this, it is certainly safe to say that, practically speaking, certain kinds of extreme injustice, like torture, tend to come back and psychically haunt, indeed undo, those who committed them. For an interesting account of the Socratic point, see Thomas Brickhouse and Nicholas Smith, “Socrates on How Wrongdoing Damages the Soul,” *The Journal of Ethics* 11(4) (2007): 337–56.

[14] For an overview, see Nir Eisikovits, “Transitional Justice,” in Edward Zalta (ed.), *The Stanford Encyclopedia of Philosophy* (Fall 2017 Edition). https://plato.stanford.edu/archives/fall2017/entries/justice-transitional.

The most rigorous philosophical account of the nature of transitional justice and the moral trade-offs involved can be found in Colleen Murphy, *The Conceptual Foundations of Transitional Justice* (New York: Cambridge University Press, 2017).

Now the operation of all transitional institutions involves moral trade-offs: Truth commissions sometimes forgo punishing the guilty if the latter cooperates with the commission's work; war crimes trials have occasionally violated the principle of legality by criminalizing behavior that was not originally illegal; lustration policies can punish those who did not have a choice about collaborating with the previous regime. Such moral trade-offs require moral justifications, and the literature on transitional justice struggles with what the best justification might be.[15]

The South African TRC, with its controversial amnesty for truth clause (perpetrators who went before the commission, and were determined to fully disclose their crimes, received immunity from prosecution), raised the question of how to justify such trade-offs with special intensity. How, to paraphrase Michael Ignatieff's memorable question, can a country allow a multitude of "murderers to get away with murder" as the inaugural act of its democracy?[16] The attempt to answer that question has spawned many scholarly attempts offering potential justifications. These included the claim that the trade-offs are justified because truth commissions promote deliberative democracy,[17] that truth commissions, by focusing on victims, recognize them as agents with stories worth telling,[18] that these bodies are better than trials at documenting past atrocities,[19] and most famously, that such bodies have the potential to promote political forgiveness and wind down the cyclical dynamics of offense and retaliation.[20]

As I have argued elsewhere,[21] each of these justifications suffers from shortcomings. The deliberative democracy justification presupposes that truth commissions must be held to the standards of justification that we

[15] For a classic collection that takes up the question of how best to justify such trade-offs, see Robert Rotberg and Dennis Thompson (eds.) *Truth v. Justice: The Morality of Truth Commissions* (Princeton, NJ: Princeton University Press, 2000). For further discussions of how to justify truth vs. justice trade-offs, see Eisikovits, *Sympathizing with the Enemy*, and Nir Eisikovits, "Peace versus Justice in Transitional Settings," *Quinnipiac Law Review* 32(3) (2014): 221–36.

[16] See Ignatieff's reportage about the TRC for The New Yorker magazine. He writes: "The reality is that, when you trade amnesty for truth, murderers get away with murder." Michael Ignatieff, "Digging Up the Dead," *The New Yorker* (November 10, 1997).

[17] Amy Gutmann and Dennis Thompson, "The Moral Foundations of Truth Commissions," in Robert Rotberg and Dennis Thompson (eds.), *Truth v. Justice: The Morality of Truth Commissions* (Princeton, NJ: Princeton University Press, 2000).

[18] Elizabeth Kiss, "Moral Ambition within and beyond Political Constraints: Reflections on Restorative Justice," in Robert Rotberg and Dennis Thompson (eds.), *Truth v. Justice: The Morality of Truth Commissions* (Princeton, NJ: Princeton University Press, 2000).

[19] Ibid.

[20] This was, in essence, Archbishop Tutu's justification for the moral compromises the commission's work involved. See Desmond Tutu, *No Future without Forgiveness* (New York: Image/Doubleday, 2000).

[21] Eisikovits, *Sympathizing with the Enemy*; Eisikovits, Peace versus Justice.

use to support institutions in settled democracies. But a truth commission is not an institution in a settled democracy. It is, rather, a body meant to help facilitate the transition to that blessed state. The justification based on recognition presupposes that the key to recognition is a sympathetic hearing rather than the criminal law's attention. The "more truth" justification ignores the fact that without the rules of evidence, the information the commissions can collect is less reliable and that commissions also have blind spots when it comes to which information they go after. Finally, the forgiveness justification struggles with successfully explaining why forgiveness should be made a subject of policy (rather than left to individuals), whether its rationale applies to conflicts between non-Christian adversaries (forgiveness doesn't have the same status in all religions), and whether resentment and even vindictiveness don't signify that the person who feels them has an appropriate degree of self-respect, which is prompting them to be resentful and angry.

More broadly, political transitions are characterized by a tension between the desire to do justice for the past and the need to preserve stability, maintain bureaucratic competence, and attract foreign investment. This tension means that transitional societies are very often caught in dirty hands dilemmas. These so-called "peace versus justice" quandaries inform both the choice of transitional instrument and its design. Thus, South Africa landed on the TRC in large part because the alternative, a war crimes tribunal to adjudicate the crimes of apartheid officials, as many members of the African National Congress demanded, would have likely resulted in civil war.

The need to address the toxic results of historical humiliation can serve as a guide in selecting and designing transitional tools, and it can also serve as a justification for the moral trade-offs these choices require. Authoritarian regimes are humiliating by definition. They use people as pawns in achieving a grand political vision or, more often, in realizing a ruler's grand vision of themselves. Such regimes care nothing about citizens' self-understanding. In addition to the loss, injury, and material misery they inflict, these forms of government also create a legacy of humiliation. Any account of life under Stalin, Saddam Hussein, or Mao will invariably describe the daily humiliations involved in not being able to protect one's family, not being able to formulate one's own life plans, having to betray one's neighbors, having to pretend to believe in what one does not believe in.

To truly make a break from an authoritarian past, transitional institutions must aim at reversing these humiliations – by creating opportunities for people to tell life stories that were previously suppressed, and by creating the educational and economic conditions for them to formulate a life plan and move to realize that plan. Humiliation, as we showed in the previous two chapters, is a profoundly dangerous psychological disposition – dangerous to the person feeling it, as it does away with their sense of agency, hope, and trust in the world and (as a result) politically dangerous. A transitional institution that tries to address this legacy by, for example, creating a public reparations program or a national storytelling project can be justified even if those programs violate some important principles that are considered sacrosanct in a settled, functioning liberal democracy (such as making people who are not guilty of causing harms participate in the payment of damages and airing inflammatory information that is not subjected to the rules of evidence). Stated differently, humiliation is a hallmark of authoritarian governance; it's baked into the very concept of authoritarianism. The feeling is both personally and politically toxic. If a process of transition is going to leave authoritarianism behind, that key element of authoritarianism needs to be addressed. And addressing it institutionally (for example, through a truth commission or lustration policies) may be worth it even if there's a dear price to pay (in our fidelity to important democratic principles).

A note of caution is in order here. Transitional justice policies can and often do make a mess of things. If the instruments of transition are deployed carelessly or too broadly, they can create a dangerous sense of humiliation among those removed from power. That sense will be especially potent for people who feel they have been collectively punished for actions they were not completely free to avoid. De-baathification and the disbanding of the Iraqi army – both badly planned instruments of lustration – are examples of how transitional justice can go terribly wrong. The humiliation they created amidst many Iraqi Sunnis who were left, overnight, without their professional dignity and without a way to support their families, was expertly exploited by Al-Qaeda in Iraq.

The catastrophic failure of lustration in Iraq is an important reminder that transitional justice measures, if calibrated incorrectly, risk extinguishing one form of humiliation at the cost of igniting another. This is particularly true considering that, as we've noted, humiliation can be idiosyncratic. People don't only get humiliated by what is objectively humiliating – they also get humiliated by events that give them little or no claim on the feeling. This is an irreducible problem with taking political humiliation seriously. Officials in a country making a transition out of authoritarian rule, who recognize the toxic role humiliation can play, must be prepared for the fact that any transitional solution they land on, even one meant to address past humiliations, has the potential to humiliate some of the stakeholders. Such officials will then face a variation of the dirty hands problem – they will have to balance the moral obligation to address normatively justified claims of humiliation made by those who have been victimized against the normatively questionable but politically explosive ones made by those removed from power.

It's important to note a striking asymmetry in what transitional justice practices are set up to handle. While transitional institutions can and often do address the root causes of humiliation, along the lines just described, they don't take up the other psychological disposition we examine in this book – namely martial glory. Bodies like war crimes tribunals, truth commissions, and lustration policies tend to ignore the glorification of war and warriors as a root cause of conflict.

At some level that is both appropriate and understandable. Even though these bodies can't always adhere to the precepts of the rule of law (for example, the Nuremberg trials are often accused of violating the principle of legality which insists that there can't be a crime without the violation of a preexisting, clearly published prohibition; truth commissions make determinations that don't always pass muster under the regular rules of evidence in a criminal trial[22]), there's pressure on them to at least look like they are trying. In that context, focusing on the way in which promises of future glory precipitated a conflict is too nebulous; it strays too far from the legalist expectation to focus on specific incidents that happened in particular times and places. Furthermore, because resources are limited during political transitions, because transitional institutions must compete with other civil society actors for funding during the process of state building, these bodies understandably focus on the most concrete instances of past wrongdoing rather than on more nebulous questions such as the dangers of martial glorification. Accounting for specific instances of past wrongdoing can show immediate political results and convey a political message in a way that investigating who glorified war and why can't.

Not only do these bodies tend to gloss over the role of glory in the run-up to war, the process of holding perpetrators accountable may actually increase the degree to which supporters glorify them. In other words, the very act of putting someone in front of a war crimes tribunal or truth commission has the potential to make more of a martyr out of them in the eyes of those still loyal to the previous regime or those who feel betrayed and left behind by the transitional process. The popularity of Serbian

[22] For an overview of how transitional institutions struggle to meet rule of law requirements, see Nir Eisikovits, Transitional Justice.

leader Slobodan Milosevic rose in his homeland as supporters watched him defend himself before the International Criminal Tribunal for Yugoslavia. As The New York Times reported after his death: "Milosevic fell from popular grace after his ouster in 2000, when his popularity ratings dropped to no higher than 10 percent, according to Strategic Marketing Research, a polling company that monitors the popularity ratings of Serbian politicians. But the sight of Milosevic in televised proceedings single-handedly taking on the Hague tribunal, who many Serbs believe is biased against them, increased those ratings to between 20 and 23 percent."[23]

The same dynamics held for Bosnian Serb leader Radovan Karadzic who stood trial for genocide and other war crimes at the Hague several years after Milosevic's death. Karadzic who was charged, among other things, with orchestrating the Srebrenica massacre in which thousands of Muslim Bosnians were murdered, was seen as a martyr by many supporters who thought the proceedings against him were biased. "Radovan Karadzic will always remain ... a hero who made an immense contribution to the creation of Republika Srpska and the defense of the Serb people," Bozica Zivkovic Rajilic, president of the Bosnian Serb Association of Women Victims of the War, told Reuters.[24] A few days before the tribunal at the Hague was supposed to read out Karadzic's verdict, Bosnian Serb officials named a student dorm near Sarajevo after him. "We dedicated this place to the man who undoubtedly set the foundation of Republika Srpska – Radovan Karadzic, the first president of this republic," the current president, Milorad Dodik, said at the unveiling of the plaque honoring Karadzic.[25]

And yet these dynamics might also work in a different direction. The process of appearing before or otherwise being vetted by a transitional body might also humanize a once-vaunted subject, complicate the way they are perceived, and replace the heroic lens with a more nuanced, tragic one. In October 1999, Robert McBride, a famous fighter for the African National Congress' military wing, the Spear of the Nation, appeared before the South African TRC and testified about his role in the African National Congress' bombing of the Why Not bar in Durban in 1986. The bombing killed three young women and injured scores of other patrons. McBride, testifying before a packed room that included the relatives of the

[23] www.nytimes.com/2006/03/13/world/europe/13iht-balkans.html.
[24] www.reuters.com/article/uk-warcrimes-karadzic-serbs-idUKKCN0WP167.
[25] www.theguardian.com/world/2016/mar/21/student-dorm-named-after-war-crimes-suspect-radovan-karadzic.

victims, was complying with the "Truth for Amnesty" provisions of the TRC's mandate. Interestingly, McBride applied for amnesty even though he didn't have to, as he had been released from prison before the beginning of the TRC process, as part of the negotiations between the apartheid government and the resistance organizations.

Like many others in the armed resistance to apartheid, he resented the equivalence the amnesty hearings created between the African National Congress and apartheid operatives – both of which had to testify before the commission. And yet McBride chose to come before the commission because he saw it as the exclusive institutional vehicle for moving beyond apartheid and was aware of the symbolic weight his testimony would carry. The impression one gets from listening to his testimony is not of an unrepentant war hero but of someone tortured by the act of violence he participated in and who acted because he couldn't find another outlet for the oppression he was subjected to under apartheid.

Asked what persuaded him to join the armed resistance, McBride replies: "Sir, I mean, I think all of us here in this place know what persuaded us to fight against oppression. I mean one doesn't have to be a rocket scientist, excuse my language. I explained to you that I was an oppressed person and that is the reason why I wanted to help the movement." McBride then continues to offer the following: "I particularly want to speak to the families of the people whose deaths I caused. I am truly sorry for causing the deaths of your loved ones. I had nothing personal against them. It was in a quest for my own freedom and in a quest to unshackle myself from the Apartheid system, that I brought about the death of your loved ones. For this I am sorry."[26]

All of this takes place before the family members of the victims, some of whom are glaring at McBride in open hostility. Listening to McBride, scanning the audience, knowing the horror that unfolded at the bar, learning about McBride's background, it's easy to put oneself in the shoes of someone attending this hearing and imagine how thoughts of martial glory are replaced by a visceral understanding of the tragedy of war, what the World War I poets called "the pity." If that's true, the transitional process here serves to demystify participation in war – to strip it of the abstract trappings of martial glory and put its full human cost on display.

[26] The quotes from the TRC hearing are taken from an excellent documentary by Frances Reid and Deborah Hoffmann, *Long Night's Journey into Day* (Iris Films, 2001). For a more comprehensive account of McBride's role in the African National Congress' armed wing, his prison sentence, the amnesty hearings, and his controversial post-apartheid political career, see Bryan Rostron, *Robert McBride: The Struggle Continues* (Cape Town, South Africa: Tafelberg, 2019).

6.3 Appeasement

Appeasement is the surrender of rights to force – giving in to the unjustified, illegitimate demands of another party (sometimes, but not always, a stronger party than oneself) in order to prevent or put off a violent confrontation. The paradigm case here is the 1938 Munich Agreement, in which the UK under Chamberlain, and France under Daladier, decided to forgo their pact with Czechoslovakia, and allow Hitler to annex the Sudetenland region of that country in return for a promise that the Germans would otherwise keep the peace. Chamberlain's speech upon returning to the UK, as he waived his signed agreement with Hitler in front of a cheering crowd ("[T]his morning I had another talk with the German Chancellor, Herr Hitler, and here is the paper which bears his name upon it as well as mine."[27]) has, fairly or not, become synonymous with shortsightedness and even cowardice in international relations.

The specter of Munich creeps up every time one country – from weakness, fear, cynicism, exhaustion, or some combination of these – allows another to unfairly exercise power. Those who propose that the Ukrainians cede territory to Russia in order to bring Putin's aggressive war to an end are accused of appeasement.[28] The Israelis accused the Obama administration of appeasing Iran, arguing that the nuclear deal signed with that country gave it free rein to plan and sponsor nonnuclear terrorism.[29] President Trump was accused of appeasing the Russians when he ignored the US intelligence reports about Putin's meddling in the 2016 elections.[30] President Biden was accused of appeasing the Saudis when he met with Crown Prince Mohammed bin Salman in an effort to control global oil prices, in spite of the alleged royal involvement in the murder of dissident Saudi journalist, Jamal Khashoggi.[31]

But why is appeasement perceived as such a grievous misstep in international relations? After all, there are certainly times when it can succeed (unlike in the case of Munich) in preventing war. Some argue that appeasement is deplorable because it emboldens the appeased party, driving it to further acts of aggression. Others add that appeasement

[27] Video of that brief is available here: www.youtube.com/watch?v=SetNFqcayeA.
[28] www.wsj.com/articles/ukraine-slams-idea-of-exchanging-land-for-peace-11653558912.
[29] www.ynetnews.com/articles/0,7340,L-4679591,00.html.
[30] https://carnegieendowment.org/2018/07/16/trump-s-appeasement-of-putin-undermines-u.s.-diplomatic-posture-pub-76844.
[31] www.dallasnews.com/opinion/editorials/2022/11/19/president-bidens-shameful-saudi-appeasement/.

undermines international norms and encourages potential aggressors watching from the sidelines to try their luck.

While these considerations are weighty, it seems that appeasement is a particularly grievous choice of policy because it involves self-imposed humiliation. By kowtowing to those who threaten violence, the appeaser communicates that mere survival or avoiding war is more important to them than their self-respect, than their view of what is good and what is bad, than their very understanding of themselves. In short, to appease is to humiliate oneself; it is to signal that one's fear crowds out any nobler sentiments and beliefs. The decision to appease is often tragic, given that the alternative courses of action can lead to disaster. Moral censure from armchair critics insulated from risk easily misses these tragic overtones. And yet, there can be little doubt that the results of appeasement are humiliating. And there is something particularly troubling – one detects a special psychological insult – in bringing such humiliation upon oneself. Conversely, the refusal to appease garners admiration: Part of the respect we feel for the Finns' conduct in the Winter War of 1940 against the Soviets, stems from the fact that they refused to appease the Russians and rejected their humiliating demands for ceding a large part of their territory. Finland ended up losing that piece of land after a few months of desperate fighting, but it earned a great deal of respect in the process. More importantly, the Soviets did not end up swallowing Finland after World War II, and the Finnish resistance set the precedent that superior power could be effectively resisted (the Ukrainians are arguably reiterating that lesson in the current war).

In his 1939 defense of the Munich Agreement, Gerald Vann tried to provide a normative principle to guide permissible cases of appeasement. Michael Walzer calls Vann's formulation the "Munich Principle." Here's Vann's defense: "If a nation finds itself called upon to defend another nation which is unjustly attacked and to which it is bound by treaty, then it is bound to fulfill its obligations ... It may, however, be its right, and even its duty, to try to persuade the victim of aggression to avoid the ultimate evil of a general conflict by agreeing to terms less favorable than those which it can claim in justice ... provided always that such a surrender of rights would not mean in fact a surrender once and for all to the rule of violence."[32] A complete surrender to the "rule of violence" is, in so many words, a surrender to humiliation – to circumstances under

[32] Vann as quoted in Michael Walzer, *Just and Unjust Wars* (New York: Basic Books, 2015): chapter 4.

which one would be done to, looked through, disposed of. So, for Vann, appeasement is acceptable only if it does not involve complete humiliation. Rephrasing him, we might suggest the following: Since what is so wrong about appeasement is that it includes a throwing away of national pride, identity, and projects – since it involves a national humiliation – it may be acceptable if the humiliation is temporary or partial, if we only give up on ourselves for a short time or if we only give up on part of our plans. Under those circumstances, perhaps there's an argument for appeasement, if the upshot is avoiding a terrible war.

Here is Keynes thinking through that kind of calculation in a letter to his friend, Kingsley Amis, as he reflects on the Munich Agreement: "One's own state of mind at the moment is painful in the way in which only a mixed state can be. Intense relief and satisfied cowardice joined with rage and indignation, plus that special emotion appropriate to the state of having been swindled; the whole nation swindled as never in its history."[33] Keynes thought that the long-term result of the agreement may be justified even if it were the product of bad judgment and weak character: "It is not certain that the present settlement may not be a good thing in the long run ... This time they may, by historic luck, have carried through a necessary thing which decent men could not have accomplished.[34] In his ambivalent letter to Amis, Keynes is trying to parse this duality – to weigh the damage of what was still considered at the time a limited national humiliation with avoiding a war that, in his estimation, the UK was not ready to fight.

And here is Clement Attlee, even more tortured than Keynes, explicitly referencing the question of national humiliation as he spoke in parliament on October 3, 1938: "We all feel relief that war has not come this time ... we cannot, however, feel that peace has been established, but that we have nothing but an armistice in a state of war. We have been unable to go in for carefree rejoicing. We have felt that we are in the midst of a tragedy. We have felt humiliation. This has not been a victory for reason and humanity. It has been a victory for brute force ... The terms have not been terms negotiated; they have been terms laid down as ultimata. We have seen today a gallant, civilised and democratic people betrayed and handed over to a ruthless despotism. We have seen something more. We have seen

[33] Letter to Kingsley Martin, 1 October 1938, in Donald Moggridge (ed.), *The Collected Writings of John Maynard Keynes Volume CWXXVIII: Political and Literary Writings* (London, UK: Macmillan for the Royal Economic Society, 1982): pp. 122–24.

[34] Ibid.

the cause of democracy, which is, in our view, the cause of civilisation and humanity, receive a terrible defeat. I think that in the mind of every thoughtful person in this Country when he heard that this settlement had been arrived at Munich, there was a conflict. On the one hand there was enormous relief that war had been averted, at all events for the time being; on the other, there was a sense of humiliation and foreboding for the future."[35]

My aim here is not so much to figure out whether Vann's argument can be defended, or if Keynes or Attlee ultimately reached the correct conclusion. It is, rather, to show that a key element involved in thinking through appeasement, in weighing surrender to unjustified demands simply because those making them can inflict a great degree of harm, has always been the recognition that doing so is humiliating. The humiliation involved is a central cost that is weighed against the benefits that can be obtained by keeping the peace.

The discussion of appeasing Germany offers some additional insights into the psychological dynamics of political humiliation. In his magisterial account of the run-up to World War II, historian Tim Bouverie argues that appeasement was underpinned by widespread sentiments of regret and even guilt among British elites over how Germany was treated by the victorious Allies after the Great War.[36] Earlier, in Chapter 5, we noted Keynes' seething criticism of the Carthaginian peace imposed on the Germans in 1919–1920. His colleague in the British delegation to the Peace Conference, EH Carr (who would later become a prominent historian), was similarly skeptical of the Versailles Agreement. Writing for the *Fortnightly Review* in 1933, he argued that "the treaty of Versailles compelled ... [the Germans] to subscribe to the most ruthless and sweeping moral condemnations in history, and riveted on the German people for another generation the fetters of their old inferiority complex."[37] Keynes' and Carr's views quickly went mainstream. "By 1933," writes Bouverie, "there was a strong feeling of *mea culpa* in Britain. Hardly anyone of influence believed that Germany had been solely responsible for the war and there was a widespread sense of guilt about Versailles."[38] Lloyd George, who was the prime minister during the Peace Conference, was sympathetic to this view and in 1933 told a large audience that the Allies'

[35] http://academic.brooklyn.cuny.edu/history/johnson/munichdebate.htm.

[36] Tim Bouverie, *Appeasing Hitler: Chamberlain, Churchill and the Road to War* (London, UK: Vintage, 2019).

[37] Ibid., p. 45. [38] Ibid., p. 46.

punitive attitude had driven Germany into "an aggressive military dictatorship."[39] Similar attitudes were echoed in the press. Thus, *The Times* proclaimed in April 1933 that the treaty "provided every grievance the heart of a German nationalist could desire."[40]

Bouverie provides a panorama of British notables who, acting as amateur diplomats, visited Nazi Germany and came back convinced that the rise of Nazism was primarily the result of the harsh sanctions imposed on Germany. On their view, the key to defanging Hitler and his party was revising the Treaty of Versailles and allowing Germany to rearm so it could take its natural place again among Europe's great powers. Thus, Philip Kerr, the Marquess of Lothian, who was Lloyd George's private secretary during the 1919 talks, argued that the most brutal aspects of Nazism sprang from the humiliations of Versailles and that, consequently, the key to reforming the regime was undoing those indignities. After meeting with Hitler, he pronounced that there were grounds for long-lasting stability and even amity between the two nations, if the British would only treat the Germans as equals. Lord Clifford Allen of Hurtwood also visited Hitler around the same time, and came away with very similar conclusions: The Versailles Treaty and France's intransigence towards Germany were very much to blame for the rise of Nazism.[41]

And so, these attitudes, which blamed the Allies for the anger and resentment that brought Hitler to power, also underpinned appeasement: If the British were to blame for the emergence of Germany's aggressive, intolerant political stance after the war, they can't just continue to push-back against their former enemy; rather they have a moral and practical responsibility to address and pacifically manage this anger. Hence the logic of appeasement.

As Bouverie puts it: "[T]he sense that the allies were to blame for the Nazis was critical to the mentality from which appeasement developed. If Britain and France had 'created' National Socialism then, logically, they could 'appease' it by redressing the grievances on which it had prospered."[42] We need not take a position on the long-standing argument between historians about the connection between Versailles and the rise of Nazism. Bouverie is certainly skeptical of the link. What *is* of interest to us is the dynamic that this historical account illustrates – that those who are involved in humiliating another (or think they are implicated in such a humiliation) can come to feel a degree of collective responsibility and guilt

[39] Ibid. [40] Ibid., p. 47. [41] Ibid., pp. 48–50. [42] Ibid., p. 47.

for it, that these feelings may motivate them to push for redressing that past humiliation, and that this attitude can inform foreign policy. In this case, the sense of responsibility and regret among British elites for Versailles (justified or not) underpinned an accommodationist and appeasing stance towards Germany. If there is something humiliating about appeasing one's foe, as we have just argued, if part of what is morally problematic about appeasement is that it involves a self-humiliation, there is a stark dynamic at play here: Humiliating a rival can set up the plausibility of being humiliated by them, as if in self-punishment, for the misdeeds of the past. In a poem about the outbreak of World War II, Auden famously writes, "I and the public know/What all schoolchildren learn/ Those to whom evil is done/ Do evil in return."[43] But the background story of appeasement laid out by Bouverie takes the claim a few steps forward: The British elites did not suffice with anticipating that those whom they wronged would want to wrong them in return; they felt a collective sense of responsibility and even guilt for Germany's plight and these sentiments led them to take a lenient view of German aggression.

Before we dismiss this episode as a historical curiosity, it's worth pointing out its echoes in the current war in Ukraine. A recurring argument made by both realist and progressive commentators holds that Russia's aggressive foreign policy in the last fifteen years or so is animated by how it was humiliated by the West when the Soviet Union fell apart and, specifically, because the United States violated promises to limit NATO expansion eastward. These sleights, so the argument goes, explain Russia's takeover of Crimea and later its full-scale invasion of Ukraine, and limit the case for American intervention against Russia. By far the most famous version of this argument has been made, for years, by the prominent realist scholar John Mearsheimer. "The United States and its European allies share most of the responsibility for the [Ukraine] crisis," he wrote in a 2014 essay titled *Why the Ukraine Crisis is the West's Fault.* "The taproot of the trouble is NATO enlargement, the central element of a larger strategy to move Ukraine out of Russia's orbit and integrate it into the West. At the same time, the EU's expansion eastward and the West's backing of the pro-democracy movement in Ukraine—beginning with the Orange Revolution in 2004—were critical elements, too. Since the mid-1990s, Russian leaders have adamantly opposed NATO enlargement and in

[43] Wystan Hugh Auden, "September 1, 1939." https://poets.org/poem/september-1-1939.

recent years, they have made it clear that they would not stand by while their strategically important neighbor turned into a Western bastion."[44]

Mearsheimer is hardly alone in this position. The Cato Institute's Ted Galen Carpenter warned in *The Guardian* last year: "History will show that Washington's treatment of Russia in the decades following the demise of the Soviet Union was a policy blunder of epic proportions. It was entirely predictable that NATO expansion would ultimately lead to a tragic, perhaps violent, breach of relations with Moscow."[45] And here is Professor Ronald Suny, a prominent expert on Soviet history at the University of Michigan: "If you think the war in Ukraine is the work of a determined imperialist, any actions short of defeating the Russians will look like 1938 Munich-style appeasement . . . If, however, you believe that Russia has legitimate concerns about NATO expansion, then the door is open to discussion, negotiation, compromise and concessions."[46]

I won't adjudicate the validity of these claims here either. But it is interesting to point out the similarities they share with the arguments that Bouverie describes for us: International aggression is explained, if not fully justified, by past historical injustices inflicted on the aggressive actor. And being implicated in those historical injustices limits the legitimacy of one's response to the act of aggression. The structure of the argument and the nature of the dynamics involved are strikingly similar: Humiliating a geopolitical rival plants the seeds for that rival's aggressive behavior while removing the grounds for pushing back against them. The appropriate response to having been humiliated is appeasement. And if it's true that appeasement involves a self-humiliation, as we have argued, the following, troubling, syllogism completes itself: There are cases when the appropriate response to having committed a humiliating act (of injustice) is to be willing to suffer a humiliation oneself.

Finally, it's worth noting that the kind of reluctant tortured cost-benefit thinking that Keynes and Attlee were engaged in necessarily changes in the nuclear age. Put simply, nuclear powers are far more likely to be appeased

44 John Mearsheimer, "Why the Ukraine Crisis is the West's Fault: The Liberal Delusions that Provoked Putin," *Foreign Affairs* (September/October 2014). www.mearsheimer.com/wp-content/uploads/2019/06/Why-the-Ukraine-Crisis-Is.pdf.

45 Ted Galen Carpenter, "Many Predicted NATO Expansion would Lead to War. Those Warnings were Ignored," *The Guardian* (February 28, 2002). www.theguardian.com/commentisfree/2022/feb/28/nato-expansion-war-russia-ukraine.

46 Ronald Suny, "Ukraine War Follows Decades of Warnings that NATO Expansion into Eastern Europe could Provoke Russia," *The Conversation* (February 28, 2022). https://theconversation.com/ukraine-war-follows-decades-of-warnings-that-nato-expansion-into-eastern-europe-could-provoke-russia-177999.

than nonnuclear ones. Indeed, one of the main arguments for preventing additional countries from going nuclear (in addition to the obvious dangers posed by proliferation) is that possessing these weapons provides a license to bully, a degree of immunity that isn't available to conventional powers. On this argument, a nuclear-armed North Korea or Iran becomes more dangerous, not only because they can start a nuclear war, but because they can harass their neighbors more freely than before, since others would become reluctant to intervene.[47] The war in Ukraine provides a helpful example of these dynamics. Russia launched a war of aggression against a peaceful neighbor. It has killed thousands of civilians, destabilized the peace in Europe, and precipitated a worldwide economic crisis. These are the kind of circumstances that called for stopping Germany in the late 1930s and that made appeasement such a problematic choice. But the circumstances have changed and the case against appeasing a nuclear power is quantitively different. Yes, Western countries have helped the Ukrainians substantively by supplying them with weapons and intelligence. But that help has been incremental and in each instance, the Americans and others have paused to see how the Russians would respond. It took more than a year of fighting for the Ukrainians to receive tanks from the West. The missile systems handed over, such as the M142 HIMARS, were supplied on the condition that they would not be used to hit targets inside Russia.[48] The rationale behind this cautious approach is straightforward and openly stated: Ukraine's allies are worried about provoking the Russians into a nuclear response and they are, consequently, treading very carefully. That is prudent. But it also means that the role of humiliation in the international system has grown exponentially.

Recall how we characterized a humiliating foreign policy in the previous chapter: The powerful throw their weight around because they can; political decisions are made based on available force with little consideration for those subjected to them. To be humiliated in foreign policy is to be done to, disposed of. Nuclear weapons increase the capacity of those who hold them to throw their weight around with impunity (or at least with more impunity than if they didn't have them). Insofar as that's true, they make it easier for the members of the nuclear club to humiliate other nations. If Russia ultimately prevails over the Ukrainians because Western

[47] For one example of this argument, see Rand Corporation National Security Research Division, "Iran after the Bomb: How would a Nuclear-Armed Tehran Behave?" www.rand.org/content/dam/rand/pubs/research_reports/RR300/RR310/RAND_RR310.pdf.

[48] On the limitations imposed on the use of HIMARS systems, see www.reuters.com/world/europe/ukraine-hasnt-used-himars-strike-russian-territory-2022-07-08/.

actors are not willing to provide sufficient help, the international system will be defined even more sharply by bullying and humiliation. Interestingly, if that happens, it won't be the Ukrainians who are humiliated but rather their Western backers, who will prove incapable of defending the kind of international system and norms that they purport to uphold. To be clear: My claim here is primarily descriptive. Ukraine's Western allies may be correct in their reluctance to defend the international order at the cost of nuclear war. Churchill had a strong argument against Chamberlain's policy in 1938 and he's appropriately credited with the courage and foresight to make it. But that was in large part because the war that Churchill foresaw would result from pushing back against the Nazis, though awful, did not have the potential to end civilization. The anti-appeasement argument cannot be repeated today with the same kind of moral certainty. The debate in the West about how much to support Ukraine rages between progressives and realists on the one hand, and liberal internationalists on the other. For the former camp, Russia's invasion of Ukraine can be largely attributed to NATO expansion after the fall of the Soviet Union and the sense of anxiety that expansion generated in the Kremlin. According to the latter view, Putin's war, much like his engagement in Syria, has little to do with encroachment on Russia's traditional sphere of influence and everything to do with the country's national self-understanding, imperial ambitions, and economic interests – all of which are independent of what the West does.[49] These explanations do not necessarily contrast as starkly as their extreme proponents assume. Russia's motivations likely have a great deal to do with how the West has behaved since the late 1980s and with its imperial self-conception. But that isn't the point here. Rather, what matters for our purposes is that in either case the freedom to rationally formulate policy in the nuclear age has been curtailed. Liberal internationalists can't freely act on their liberal internationalism. Progressives can't freely act on their progressivism, and realists have a very narrow set of interests that they can take seriously once a credible threat of nuclear war has been introduced. An international system defined by nuclear deterrence becomes structurally humiliating, or at least more structurally humiliating, than a system that isn't. Nuclear weapons multiply the force of the powerful and curtail the freedom of principled parties to act on their beliefs. Under these circumstances, weaker actors are at increased risk of becoming subject to the whims of

[49] For a good summary of this argument, see Rajan Menon, "How to End the War in Ukraine," *Boston Review* (April 26, 2023). www.bostonreview.net/forum/how-to-end-the-war-in-ukraine/.

the strong, and third parties are forced to (at least partially) subsume their moral commitments to their interest in surviving. Both eventualities are humiliating.

6.4 Summary

Thinking about humiliation and its consequences informs various areas of political theory – even if latently. Part of the point of classical jus in bello restrictions like the requirements of proportionality and discrimination is to limit the harm we do to our enemies, so as to keep alive the possibility of future reconciliation. Indiscriminate and disproportionate harms undermine the chances of peace, among other reasons, because they are humiliating. The realist take on the same argument suggests that it is important to refrain from humiliating one's enemies not only in order to keep the chances of peace alive but also in order to prevent the tendency to humiliate others from turning inwards and becoming a domestic practice.

In the field of transitional justice, the prospect of ending the humiliations endemic to authoritarian governance can justify the compromise of liberal principles (such as retroactive criminalization and reliance on shaky evidence) that transitional policies often involve. We noted that transitional processes are typically better at addressing past humiliations than they are at accounting for the glorification of problematic policies and leaders. And yet, such tools for coming to terms with the past can, under some circumstances, help replace glorification with more sober attitudes.

Our discussion also took up the role humiliation plays in political appeasement. We argued that one of the reasons that appeasement is wrong is that it involves a self-humiliation. By deferring to those who threaten force, the appeaser communicates that survival matters more to them than their self-respect. To appease is, most often, to humiliate oneself. We also suggested that there are cases where the willingness to appease others springs from a sense that one has, in the past, unjustly humiliated them. Finally, we argued that the likelihood of appeasement and humiliation in foreign policy necessarily grows after the introduction of nuclear weapons.

PART III

Glory and Humiliation: War's Two-Stroke Engine

CHAPTER 7

War's Two-Stroke Engine

Part I of this book explained the nature of martial glory. Part II gave a theoretical account of political humiliation and its role in international relations. But how are the two – glory and political humiliation – related? In this chapter, I will argue that glory and humiliation act in conjunction with each other – indeed act on each other – like the components of a two-stroke engine – to drive war forward. After making that claim, I will, in the next chapter, illustrate these dynamics by looking at three case studies.

7.1 The Glory-Humiliation Nexus

7.1.1 Glory for Reversing Humiliation and the Nature of Political Purposes

The first facet of the relationship between humiliation and glory is straightforward: Being humiliated is so injurious psychologically and socially that ending or reversing that state results in the bestowal of glory. The sense of being ignored, done to, disposed of, displaced, or replaced, the realization that one can't act on one's idea of oneself because of another's coercion, that one has become an instrument of another's will – all dispositions that are constitutive of political humiliation as we've described it – are so damaging that reversing them becomes a key political and psychological need. And because the need is great, those who are engaged in meeting the need are robustly recognized; they win glory.

To use the terms of Periclean glory developed in Part I, ending humiliation is the kind of political purpose that carries glory in tow for those who will risk their lives for it. The removal of humiliation brings not only relief and gratitude but also super recognition.

A legacy of humiliation and the bestowal of glory work, then, in tandem: Ending political humiliation is an important political purpose or project; serving this purpose is so politically important that it is rewarded with glory, and glorifying then cements the sanctity of the purpose and makes people want to serve it even more. Hence, the need to end humiliation and the glory that comes with meeting that need generate a self-reinforcing cycle.

The mediating term we employ here – "political purpose" – requires further explication. Service to an important political purpose wins us glory, as Pericles insists, but what are political purposes, and more importantly, why are people willing to serve them, even go to war for them, in the first place? In his famous 1910 essay, "The Moral Equivalent of War," William James acknowledges that war has a profound psychological draw. It is awful, but it's not only awful. War's rigors spark the imagination in a way that few other activities can: "The horrors make the fascination," James writes, bleakly. "War is the strong life; it is life *in extremis*."[1] The moral outlook and qualities forged in military activities, he argues, are necessary for civic virtue; they are key to overcoming tendencies to self-absorption and shallowness typical of contemporary life. How is a progressive who, nevertheless, recognizes these facts about the allure of war supposed to think about bringing it to heel? For James, the answer is obvious: Find a nonmilitary context that instills the same kind of virtues and generates the same kind of excitement. Find a way, in other words, to preserve the psychological benefits of going to war in the name of a great cause, without paying its physical costs:

> I do not believe that peace either ought to be or will be permanent on this globe, unless the states pacifically organized preserve some of the old elements of army-discipline. A permanently successful peace-economy cannot be a simple pleasure-economy ... we must still subject ourselves collectively to those severities which answer to our real position upon this only partly hospitable globe ... Martial virtues must be the enduring cement; intrepidity, contempt of softness, surrender of private interest, obedience to command, must still remain the rock upon which states are built.[2]

The point of James' militarized version of pacifism is straightforward: As awful as war is, the national galvanization it offers, the sense of collective purpose, the feeling of subsuming one's personal interests to an important cause, all have claims on the human imagination that are difficult to compete with, and, importantly, carry real social benefits. He continues:

[1] Originally published as William James, "The Moral Equivalent of War," *Lecture 11 in Memories and Studies* (New York: Longmans, Green & Co, 1911): pp. 267–96, p. 269.
[2] Ibid., p. 288.

> All the qualities of a man acquire dignity when he knows that the service of the collectivity that owns him needs them. If proud of the collectivity, his own pride rises in proportion. No collectivity is like an army for nourishing such pride; but it has to be confessed that the only sentiment which the image of pacific cosmopolitan industrialism is capable of arousing in countless worthy breasts is shame at the idea of belonging to such a collectivity. It is obvious that the United States of America as they exist to-day impress a mind . . . as so much human blubber. Where is the sharpness and precipitousness, the contempt for life, whether one's own, or another's? Where is the savage 'yes' and 'no,' the unconditional duty? Where is the conscription? Where is the blood-tax? Where is anything that one feels honored by belonging to?[3]

James concedes, then, the emotive force of militarism – a force largely derived from the psychological satisfaction one gets from serving a political purpose or cause beyond oneself. He understands the corresponding weaknesses of pacifism, and yet is a committed socialist pacifist. He realizes that the way out of this conundrum is to offer people a "moral equivalent of war,"[4] a decaffeinated conception of national purpose, if you will, an alternative collective project that could fascinate and excite people equally – enlisting them perhaps in the cause of social justice, or to fight a public health crisis, or in a struggle against natural disasters – any of these could, possibly, fulfill the same function. What is needed is civic conscription: "To coal and iron mines, to freight trains, to fishing fleets in December, to dishwashing, clothes-washing, and window washing, to road-building and tunnel-making, to foundries and stoke-holes, and to the frames of skyscrapers, would our gilded youths be drafted off, according to their choice, to get the childishness knocked out of them, and to come back into society with healthier sympathies and soberer ideas."[5]

James' attempt to distill from militarism only its attachment to grand political causes has an obvious flaw: He does not seriously entertain the possibility that the substitutes he has in mind would lose the energy that animates the original, that one cannot have the psychological benefits of militarism without the violence, that, in fact, the violent "costs" are the source and condition of the benefits accruing in the first place. Military causes might, in other words, be as psychologically alluring as they are exactly because one might die defending them or kill in their name.

Several decades later, George Orwell offered a bleaker variation on James' view. In his 1940 review of Hitler's *Mein Kampf*,[6] Orwell argued

[3] Ibid., p. 286. [4] Ibid., p. 283. [5] Ibid., p. 291.

[6] George Orwell, "Review of Mein Kampf," in *George Orwell: Essays* (New York: Everyman's Library, 2002): pp. 249–50.

that what made war attractive (and not merely an agent of destruction and trauma) was the lure of great political purposes. Becoming entangled with a great cause, being subsumed by it, Orwell seems to suggest, is irresistible to us. Fascists and totalitarians never completely go out of fashion, he argued, because they understand something that decent liberals have a harder time understanding: People do not only want the material betterment of their own condition. They also want ideals larger than themselves, dangerous yet seductive political projects that they can disappear into, perhaps even sacrifice themselves to: "Hitler, because in his own joyless mind he feels it with exceptional strength, knows that human beings *don't* only want comfort, safety, short working-hours, hygiene, birth-control and, in general, common sense; they also, at least intermittently, want struggle and self-sacrifice, not to mention drums, flags and loyalty-parades. However they may be as economic theories, Fascism and Nazism are psychologically far sounder than any hedonistic conception of life. The same is probably true of Stalin's militarized version of Socialism. All three of the great dictators have enhanced their power by imposing intolerable burdens on their peoples. Whereas Socialism, and even capitalism in a more grudging way, have said to people 'I offer you a good time,' Hitler has said to them 'I offer you struggle, danger and death,' and as a result a whole nation flings itself at his feet."[7]

People are fascinated by war, Orwell tries to tell us, not simply because it is exciting; rather, it is exciting to them because it fulfills a basic need – moral as much as psychological – to go beyond oneself and to be subsumed by a great purpose or idea. War answers that need more completely and fully than the alternatives; it's more efficient in satisfying this craving for transcendence than giving to charity or joining the peace corps. Orwell's brief argument raises some serious doubts about the half-hearted optimism at the end of James' essay. After all, as any Freudian would tell James, "other outlets for heroic energy" are as useful as the original only if they discharge the same amount of energy. But what's the reason to assume that they can release the same amount of energy? If war is simply a contingent way to answer our needs for excitement and belonging, what explains its remarkable, millennia-long staying power? James argues that it doesn't have to be war that generates the special attractions, but he doesn't offer an argument to justify his claim that alternatives would do as well. Stated in psychoanalytic terms, what James offers is sublimation, and sublimation, as Freud told us, does not serve the same energetic purpose

[7] Ibid.

as the activity that it seeks to replace.[8] "The Socialist who finds his children playing with soldiers is usually upset," Orwell remarks, and then he departs from James: "... but he is never able to think of a substitute for the tin soldiers; tin pacifists somehow won't do."[9]

Machiavelli makes a related argument in an oft-overlooked passage in *The Prince*. A leader, he argues, need not worry about placing taxing demands on his subjects; after all, people get attached to political ideas as much because of what these ideas demand of them as for what the state gives them. A prince should not fret about soldiers getting injured in his service or losing their homes in the course of war because such injuries and losses will likely cement rather than sever patriotic ties. There's a cold psychological logic to this: A soldier who loses a limb in war and becomes disgruntled with the state for his loss becomes nothing more than a disabled citizen. Conversely, one who views the loss as something one has *given* to one's country is a war hero. As Machiavelli puts it, succinctly, "[T]he nature of men is such that they obligate themselves to another as much for the benefits they give as for those they receive."[10]

One can also consider Ernst Jünger's description of the draw of political causes, from his famous World War I memoir: "And so, strange as it may sound, I learned from this very four years of schooling in force and in all the fantastic extravagance of material warfare, that life has no depth of meaning except when it is pledged for an ideal, and that there are ideals in comparison with which the life of an individual and even of a people has no weight."[11]

It is probably this Orwellian/Machiavellian/Jüngerian insight that explains why war is so often described not only as an agent of terrible damage but also as an object of attraction, fascination, infatuation, and even love. Here's Jünger again, in a later edition of his memoir: "We had come from lecture halls, school desks, and factory workbenches, and over the brief weeks of training, we had bonded together into one large and enthusiastic group. Grown up in an age of security, we shared a yearning for danger, for the experience of the extraordinary. We were enraptured by

[8] See, for example, Sigmund Freud, *Civilization and its Discontents*, trans. Strachey (New York: Norton, 1961): chapter 2, p. 30. Freud writes about sublimations or "displacements": "[A]t present we can only say figuratively that such satisfactions seem 'finer and higher.' But their intensity is mild compared from that derived from the sating of crude and instinctual impulses; it does not convulse our physical being."

[9] Orwell, Review of Mein Kampf. [10] Machiavelli, *The Prince*, p. 64.

[11] Ernst Jünger, *In Stahlgewittern: Aus dem Tagebuch eines Stosstruppführers* (Leipzig, Germany: Robert Meiser, 1920), translated into English as *The Storm of Steel* (London, UK: Chatto and Windus, 1929): pp. 316–17.

war. We had set out in a rain of flowers, in a drunken atmosphere of blood and roses. Surely the war had to supply us with what we wanted; the great, the overwhelming, the hallowed experience."[12] Examples can be multiplied, from Nikolai Rostov's pronouncements about his love of his Tsar and desire to die for him in *War and Peace* to actual soldiers' reports insisting that, alongside the trauma and loss, war gave them a sense of purpose and direction. The point should be clear enough: Fighting for a great political purpose is not only a philosophically coherent proposition; it is, at least sometimes and at least for a while, a psychologically attractive one. In fact, it may very well be that our attraction to such purposes *is* what makes war "interesting." People can, and often do "fall in love," to use Thucydides' description of how the Athenians were lured in by the Sicilian expedition, with such national purposes and projects.

If this thirst for a great, overwhelming purpose or idea is part of what draws people to war, humiliation often supplies that idea with content. So great is the hold of humiliation on the political imagination that the desire to reverse it has, historically, been a go-to political purpose. The French were to fight in order to reverse the humiliation the Germans inflicted on them in the Franco-Prussian War; the Germans were to fight to reverse the humiliation the French and others imposed on them in the aftermath of the Great War; the Egyptians went to war to reverse the humiliation the Israelis visited on them in 1967. ISIS leaders wanted their caliphate because, they said, only its establishment could reverse centuries of humiliation by the infidels. The Chinese are to reassert themselves on the international stage in order to reverse their "century of humiliation" at the hands of Western powers. Putin frames his war in Ukraine and his concerns about NATO encroachment by invoking decades of humiliation by the West and particularly the Americans in the aftermath of the collapse of the Soviet Union. And so it goes. People have a genuine need for a political purpose; the reversal of political humiliation can and often has been a prime candidate for such political purpose, and that purpose is often perceived as so important that fighting for it is recognized by political glory. This recognition, this bestowal of glory, both cements the status of the political purpose as great and helps bring more people to it, as many more become hungry for recognition. Humiliation and glory thus serve as the two-stroke engine of war – the movement of the former provides momentum to the latter which then returns to reanimate the former and so on.

[12] Ernst Jünger, *Storm of Steel,* trans. Hofmann (New York: Penguin, 2004): p. 5.

7.2 Humiliation as Lost Glory

There is, however, another aspect to the relationship between humiliation and glory. If the desire to reverse political humiliation is often a cause or purpose that is rewarded with glory, it is important to notice that political humiliation is often understood or *defined as* lost glory.

In other words, the content of the sense of political humiliation is the feeling of having been glorious once and having lost that recognition to others, who are perceived as less deserving. This conception of humiliation as lost glory is key to the sense of replacement that we discussed earlier. A variation on this conception is the conviction that one has not received as much glory as one (thinks one) deserves.

China's "century of humiliation" is rooted in the sense of having been the "Middle Kingdom" – at the pinnacle of power and prestige and having lost it to "barbarians" from the West or Japan. ISIS' humiliation cashes out, at least partly, to a perception that previously "warlike" and formidable Sunni Muslims have now been subjugated by "crusaders" and Jews. In Putin's telling, Russia was once a conquering superpower that other nations trembled before and whose interests were always taken seriously by its competitors, but, after the Cold War, its imperial glory faded, and it started receiving dictates from the superficial and self-satisfied Americans. In all these cases, humiliation *is* just the loss of former glory. The political sentiment consists in having lost much-deserved recognition. The sense of forfeiting status (and the resulting nostalgia) is, in these cases, constitutive of the feeling of humiliation.

7.3 A Partial Symmetry

Though the concepts of glory and humiliation are intertwined in the ways just described, the relationship between them is not symmetrical. While it's often true that the reversal of humiliation is the political project that promises to bestow glory, while humiliation often consists in the feeling of having unjustly lost or never received the glory we deserve, glory can motivate people to go to war independently of whether they have been humiliated. We can fight for glory even if a sense of humiliation is not present. There is glory in achieving great political causes that have nothing to do with historical humiliation. The Americans glorified their "greatest generation" for fighting fascism; the Israelis glorified their paratroopers for capturing the Old City of Jerusalem in 1967; first responders were glorified in the wake of the September 11 attacks for rushing into the wreckage of the Twin Towers. In all of these cases, those glorified were so recognized because they fought for or took risks in the name of (what was believed to be) a great national purpose. And that purpose was not importantly related to feelings of national humiliation.

There is, then, a strange lopsidedness in the relationship between the ideas of glory and humiliation: While it is certainly possible to receive glory without fighting to end political humiliation, fighting to end political humiliation will almost always involve a promise of glory. It is, in fact, hard to think of a political example in which that is not the case.

7.4 A Complete Asymmetry

Putting this conceptual asymmetry between the two terms to one side, it's crucial to note a practical asymmetry in the way glory and humiliation are distributed. Historically, fighting has been an activity conceived of, planned by, and carried out by men. That has meant that military honors for fighting well, or for a worthy cause, have been primarily bestowed upon men. The costs of war, however, have been borne by both combatants and noncombatants – primarily women but also children and men beyond the age of fighting. The so-called "civilian casualty ratio," which denotes the proportion of noncombatants among war casualties, has consistently risen with the professionalization of armies, the development of military technologies, and the increase in asymmetrical and hybrid forms of warfare. One recent meta-analysis puts the ratio at 5 percent early in the nineteenth century, 15 percent in World War I, 65 percent during World War II, and over 90 percent in the wars of the 1990s.[13] Experts speaking before the UN Security Council in the spring of 2022 reiterated the 90 percent figure, and stressed the need for the Council to protect uninvolved civilians.[14]

Whether these numbers are accurate or whether the ratios are still somewhat lower,[15] there's a dramatic mismatch between who gets the recognition for going to war and who suffers from it. Put differently, when

[13] Amir Khorram-Manesh, Frederick Burkle, Krzysztof Goniewicz and Yohan Robinson, "Estimating the Number of Civilian Casualties in Modern Armed Conflicts–A Systematic Review," *Frontiers in Public Health* 9 (2021). https://doi.org/10.3389/fpubh.2021.765261.

[14] See press release from the meeting of the UN Security Council, "Ninety Per Cent of War-Time Casualties are Civilians, Speakers Stress, Pressing Security Council to Fulfil Responsibility, Protect Innocent People in Conflicts" (May 22, 2022). https://press.un.org/en/2022/sc14904.doc.htm.

[15] For a more skeptical note, see Brian Smith, "Two Dogmas of the New War Thesis," *International Journal of Military History and Historiography* 38(1) (2018): 92–120.

men go to war, they certainly have a great deal to lose. Many of them go because they are forced to and not because they were moved by a rousing Periclean or Churchillian speech. But men can also gain the social capital that comes with having fought (glory or the other status and economic advantages that accrue to veterans). The picture for women (and children, and men beyond the age of fighting) is very different; war is often an activity imposed on them, wherein they shoulder many of the costs and reap very few of the benefits.

It's important to complicate this description. In recent years, democratic societies have striven to increase the access and participation of women in military pursuits. And noncombatant family members of those who fight often take great pride in the service of their relatives. Fighting for an important cause is frequently understood by noncombatants as a communal endeavor rather than as simply the pet project of the menfolk. That is particularly true in asymmetrical wars where combatants regularly take refuge among noncombatants and use civilian cover to evade capture. And yet, in parallel, it is also hard to avoid the stark social reality that war is historically something that men *do* and that women (and children) *suffer*. As Hugo Slim, a scholar of humanitarian law, puts it, "The cultures and institutions that prepare for and deliver organized armed violence on behalf of the State or non-State armed groups are predominantly constructed by men, led by men and filled with men."[16] If war is sometimes understood as a communal undertaking, there are certainly also plenty of cultural and historical signposts that show it as an activity wherein men get to display agency and women are left with the bill. Recall our earlier discussion of how Andromache begs Hector not to go to battle in Book VI of the *Iliad*: "Husband, this courage of yours dooms you. You show no pity for your little son or your wretched wife ... Hector you are parent, brother, husband to me. Take pity on me now, and stay here on the battlements, don't make your son an orphan and your wife a widow." Though he understands what leaving Andromache behind entails for her, though he tells his wife that he shudders to think of the moment when "some bronze-clad Greek drags you away weeping," Hector insists he must carry out his duty. Anything else would be both shameful and go against his nature:

[16] Hugo Slim, "Masculinity and War—Let's Talk About It" in *Humanitarian Law & Policy* (International Committee of the Red Cross Blog, March 15, 2018). https://blogs.icrc.org/law-and-policy/2018/03/15/masculinity-and-war-let-s-talk-about-it-more/.

> If I hid from the fighting like a coward, I would be shamed before all the Trojans and their wives in their trailing robes. Nor is it my instinct, since I have striven ever to excel always in the vanguard of the battle, seeking to win great glory for my father and myself.[17]

And, for the sake of variation, here's Rudyard Kipling offering his take on the theme in the "Harp Song of the Dane Women." Though the narrator's husband seems to like domestic life, the itch of war and its excitements have a stronger hold on his imagination:

> Yet, when the signs of summer thicken,
> And the ice breaks, and the birch-buds quicken,
> Yearly you turn from our side, and sicken—
>
> Sicken again for the shouts and the slaughters.
> You steal away to the lapping waters,
> And look at your ship in her winter-quarters.
>
> You forget our mirth, and talk at the tables,
> The kine in the shed and the horse in the stables—
> To pitch her sides and go over her cables.
>
> And hence the poem's heartbreaking conclusion with its straightforward description of the differential impact of war on the genders:
>
> Ah, what is Woman that you forsake her,
> And the hearth-fire and the home-acre,
> To go with the old grey Widow-maker ?[18]

Not only do wars take an unequal physical toll on men and women,[19] they also have a disproportionate impact on their mental health – due to the fear of (and aftermath of) sexual violence, the stress of worrying about a spouse off fighting while one is taking care of children alone, suffering from food insecurity, and other worries.[20] In fact, to the extent that humiliation is one of the mental health costs of war, it is safe to say that women are primarily the ones who pay that cost – both because they have to contend with being done to and acted upon by the (primarily male) agents of war and because they often lack the agentic outlet available to

[17] Homer, *Iliad*, book VI.

[18] Rudyard Kipling, "Harp Song of the Dane Women." www.poetryfoundation.org/poems/47691/harp-song-of-the-dane-women.

[19] For a good overview, see Hattie Phelps, "The Disproportionate Effects of War and Conflict on Women and Girls," *The Georgetown Journal of Gender and the Law* 23(3) (2003).

[20] See, for example, Murthy Srinivasa and Rashmi Lakshminarayana, "Mental Health Consequences of War: A Brief Review of Research Findings," *World Psychiatry: Official Journal of the World Psychiatric Association* 5(1) (2006): 25–30.

men, who can at least try to offset the psychological and material injuries of war by fighting back and can be buoyed by the public recognition of doing so.

Let's put this last point as clearly as possible: The data on the differential impact of war on men and women suggest a stark syllogism. If war is an activity conceived, planned, and executed by men, and if that activity has lopsided costs for women, and if one of the key features of political humiliation is the sense of being done to, being disposed of, having projects carried out at one's expense without regard to one's views on the matter, then war is inherently humiliating for women, but not for men. Yes, the statement needs qualification. Certainly, plenty of men become unwilling tokens in a game of war played over their heads. And yes, women do increasingly participate in planning and fighting wars. And yes, some women who do not plan or fight in wars are nevertheless personally committed to the causes being fought for and accept the sacrifices involved. And yet, the deep history is hard to ignore: If the figures about the civilian casualty ratio are even close to being accurate, war is essentially and increasingly something that men do to noncombatants, disproportionally to women, without asking them. The gender imbalance between how the rewards and costs of war are distributed suggests that war, by its nature, is humiliating for women.

7.5 Glory, Humiliation, and Private Life

We've discussed the relationship between humiliation and glory: We noted that reversing national humiliation is often the political cause that a country rewards with glory. We explored how humiliation is often experienced as glory denied – as having lost a special kind of standing or recognition one used to have. We also noted the asymmetry in the distribution of glory and humiliation in the history of war. In wrapping up this chapter, it's worth raising one last question about the comingling of glory and humiliation, namely – what might a preoccupation with these two dispositions mean for our private lives?

People who are consumed by the humiliation and glory cycle, those moved to reverse political humiliation (and win the recognition that comes with the reversal), are, by definition, living lives in which the private is subordinated to the public. To gravitate towards the glory/humiliation dyad is not only to ascend from the private, as Arendt saw it, it is also to exact costs from the private. These pursuits carry a price for those who make up the intimate circle of the agent in question. And those who pay that price – spouses, children, elderly parents – are typically not asked if they want to pay it. The ancient Greeks may have dismissed this intimate milieu; they may have viewed it as the realm of necessity, which those who are primarily interested in political action must rise above. But we are not ancient Greeks. If the liberal commitment to the fundamental value of individual lives means anything at all, then the private costs of a life lived in the glory-humiliation nexus must be tallied.

The person who would perish to reverse a political humiliation is in many ways impressive. But their virtue can also be a hard, dogmatic one, and that is particularly true when the lure of the great cause makes them look through the people immediately around them. Recall the exchange between Hector and Andromache just quoted: It's not that Hector does not care about his wife and child; it's not that he's blind to the costs the war would impose on them. In fact, the parting scene between the husband and wife is one of the most poignant and heartbreaking in the *Iliad*. But in Hector's world, defined as it is by the need to avoid shame and bring glory to his royal family, the private (what happens to his wife and infant son) clearly takes second place.

The same preoccupation with the priority of humiliation and glory is prominent in Sophocles' *Ajax*. But if Hector is tortured by his choice, Ajax displays few qualms. After Achilles dies in the Trojan War, the question arises who should inherit his armor? Ajax thinks he has a good claim to the honor because, as far as he's concerned, he's clearly the next best warrior after the dead hero. The two kings leading the Greek expedition, Agamemnon and Menelaus, disagree and decide to give the armor to Odysseus instead. Enraged and humiliated by this snub, Ajax decides to kill all three. But he is tricked by the goddess Athena, who makes him delusional. Ajax slaughters herds of cattle, all the while convinced that he is killing his foes. When he realizes what happened, the great fighter is so humiliated that he decides to take his own life. Ajax is first humiliated by not getting the recognition he thought he deserved, then by the trick Athena played on him, and, finally, by the possibility of becoming a laughingstock. As he describes it, his sense of humiliation involves a terrible gap between how he perceives himself and how others will now see him: "[H]ere I am the bold, the valiant, unflinching in the shock of war – a terrible threat to the unsuspecting beasts. Oh what a mockery I have come to. What indignity."[21] The conclusion, for Ajax, is bleak. This

[21] Sophocles, *Ajax*, trans. Moore, in David Grene and Richard Lattimore (eds.), *Sophocles II* (Chicago, IL: University of Chicago Press, 2013): line 365.

humiliation has undone his very identity: "[A]ll that I was has perished with these poor creatures here."[22]

Ajax's terrible distress illustrates some of the theoretical points we've made earlier about the phenomenology of humiliation: It's an example of humiliation understood as lost or denied glory (Ajax was humiliated by the withholding of recognition; he could not come to terms with being passed over for Achilles' armor). It also illustrates the porous lines between humiliation and embarrassment: The public airing of a gap between how one perceives oneself and how others perceive them is, strictly speaking, embarrassing rather than humiliating, but this conceptual parsing hardly mattered to Ajax.

When she begins to suspect that Ajax might kill himself because of this humiliation, Tecmessa, a war captive and the mother of Ajax's son, urges him to remember her and their child, and to prefer their well-being to his injured honor.[23] Much like Andromache, she fears that if Ajax leaves them behind, they will be abused by his enemies. She urges Ajax to remember that she has been kind and supportive of him: "O my lord, remember even me. A man ought to remember if he has perhaps experienced some pleasure. Kindness it is that brings forth kindness always; but when a man ignores good done to him and lets the recollection of it slip away, how shall I any longer call him noble?"[24]

Tecmessa offers us an alternative notion of nobility to the heroic one espoused by Hector and Ajax: It consists in fidelity to those in your immediate circle, in returning a private kindness shown to you. She urges Ajax to prefer decency to glory. This may be a hopeless request in the world of the ancient Greeks, preoccupied as they were with honor and shame, but I am not sure about that. After all, Tecmessa and her complaint are compellingly rendered by a classical Greek playwright. The Arendtian view according to which the Greeks had a degree of contempt towards "the private" is challenged by the fact that both Homer and Sophocles give Andromache and Tecmessa, representatives of this supposedly contemptuous "private" realm, powerful, moving voices.[25] Be that as it may, a

[22] Ibid., line 405. [23] Ibid., line 510. [24] Ibid., lines 520–25.

[25] Here is another piece of evidence to buttress this "anti-heroic" case: One notes that Odysseus, one of the targets of Ajax's rage, is not rigidly moved by a concern for glory and humiliation. He would rather that Athena did not humiliate Ajax in the first place, and he is genuinely grieved at the prospect of Ajax's body being defiled after death. As far as Odysseus is concerned, Ajax – foe or not – deserves recognition for his military attainment and nobody should be humiliated in death. Ajax, Odysseus notes, is "my enemy it's true. But he was noble" (Ibid., line 1355). "His greatness," he adds, "weighs more than my hate with me" (Ibid.). Sophocles' Odysseus rejects the rigid friend/enemy distinction that Athena tries to impose on him. When it becomes clear to the goddess that,

twenty-first-century reader senses, upon reading these works, that a preoccupation with glory and humiliation is in tension with tending to the private, that a person who has a family should not love glory and be offended by humiliation as much as Ajax is.

But back to the play. Ajax is not impressed with Tecmessa's plea. Just before he kills himself, he asks Zeus to make sure that his brother finds his body first, rather than an enemy who would mutilate it: "[S]end some messenger . . . to bring the evil news about me first to Teucer, so that he may be the first to lift me up when I have fallen in this fresh stained sword. I would not have some enemy spy me out and cast me for a prize for the birds and dogs."[26] What Ajax doesn't think about is what actually happens – Tecmessa is the one who finds him first. Even as he is about to perish, Ajax cannot spare a thought for those who make up his private circle; he can't bring himself to think of anything but the need to avoid humiliation.

The most striking literary illustration of this tension between a preoccupation with humiliation on the one hand, and the capacity to lead a satisfying private life on the other, is found in Heinrich von Kleist's novella, *Michael Kohlhaas*. Kohlhaas, a respectable and wealthy horse trader from sixteenth-century Brandenburg (in the northeast of today's Germany), sets out on a trip to sell some of his horses. Upon crossing to a neighboring province, he is detained by the officials of a local nobleman who demand a toll and permit if he is to pass through their territory. The demands are unfamiliar and seem suspicious to the horse dealer. But he leaves two horses behind as collateral and is permitted to continue his trip. Soon thereafter Kohlhaas figures out the requirements were, indeed, spurious but when he returns to collect his horses he discovers that they have been terribly maltreated and that the groom he left to care for them has been badly beaten. Even though he is upset, Kohlhaas maintains his faith in the legal system and sues the nobleman who wronged him. But his suit is thrown out because of his adversary's political connections.

Humiliated, our protagonist cannot let the insult go. Even though he has a thriving business and a satisfying family life, even though this setback destroys neither, Kohlhaas is shaken to his core by the fact that this could

in spite of his enmity, he has no interest in humiliating Ajax, she is puzzled: "To mock foes, is not that the sweetest mockery?" (Ibid., line 77). Odysseus, portrayed in the Odyssey as *multitropos* or the "man of many ways," displays some of his signature flexibility here. More than anything he seems to have some empathy for Ajax and that empathy seems the very antidote to the single-minded unbending orientation that characterizes his dead foe.

[26] Ibid., line 825.

happen to him. He can "no longer take any pleasure either in his horse breeding or his house and farm, hardly even in his wife and children,"[27] and he gradually becomes obsessed with getting justice and reversing the humiliation he has suffered. He sells his properties at a loss to finance his campaign. He tells the surprised bailiff who is about to buy them that "Kohlhaasenbruck (the family farm) was, after all, not the world" and then he notes that "there might be purposes in life compared to which that of being a good father to his family was an inferior and unworthy one." He adds that "his soul aimed at great things, about which he (the bailiff) would perhaps be hearing shortly."[28]

After this ominous statement of priorities, Kohlhaas, increasingly consumed by his desire for justice, begins to shed his identity as a family man. He tells his wife, Lisbeth, "I will not go on living in a country where they don't protect me in my rights. I'd rather be a dog if people are going to kick me than a man!"[29] The repeated use of the first-person pronoun is striking here as it indicates that Kohlhaas is beginning to forget that he does not live alone. He is a husband and a father. The question of whether a family man can afford to love justice or hate humiliation to the exclusion of everything else does not occur to him. Lisbeth, in a desperate attempt to help her husband and lay the problem to rest, suggests that she may go to the capital city and bring his case to the attention of the elector of Saxony. One of the elector's senior officials had courted her long ago and she thinks she can make use of that old relationship to get through. Kohlhaas agrees but Lisbeth is killed, under murky circumstances, in her attempt to deliver her husband's petition. And it is here that Kohlhaas' preoccupation turns into a violent vendetta that eventually embroils the entire country. He gradually gathers an army of malcontents and rebels and declares himself to be fighting what amounts to a holy war against the corrupt officials who have denied him justice. The end of the novella finds Kohlhaas about to be executed. He is in possession of a document that has sensitive information which is crucial for a key political figure. The horse dealer is offered his freedom in return for surrendering the document. The stakes could not be clearer. Kohlhaas could help the authorities and go back to what's left of his family. His children, already orphaned from their mother, would at least have a father. But he won't hear of it. By holding on to the document, he can hurt his foes. Kohlhaas loves "the

[27] Heinrich von Kleist, *Michael Kohlhaas*, trans. Greenberg (Hoboken, NJ: Melville House Publishing, 2005): p. 25.

[28] Ibid., p. 27. [29] Ibid., p. 30.

power given him to wound his enemy mortally in the heel at the very moment that it was treading him in the dust."[30] An old woman visits him in jail and urges him to hand over the document. "Not for the world, Granny, not for the world," he responds. The old lady (who bears an uncanny resemblance to Kohlhaas' dead wife), puts his youngest child on her lap and says, "not for the world, Kohlhaas the horse dealer, but for this pretty little fair-headed boy."[31]

Kohlhaas' trajectory crystalizes the stakes involved in pursuing the great causes. To be drawn into them is to be drawn away from one's private life and private concerns. It is, as Kohlhaas says explicitly, to become convinced that those concerns are small. The lover of the great political purposes is drawn in by "the world." He becomes accustomed to thinking in universal terms. When the old lady tells Kohlhaas that the world has nothing to do with his choice and that he should consider giving the document up for a very particular boy (his own), he does not quite know how to respond. It's as if he is no longer capable of thinking at this more intimate scale.

Hector, Ajax, and Michael Kohlhaas represent a type of personality. In each case, the emphasis is a bit different – Hector is preoccupied with what he sees as his princely duties, Ajax can't bear how the goddess has humiliated him, and Kohlhaas is obsessed with revenge as the only means to reverse his humiliation by the authorities. But all three men assign far greater weight to these causes than they do to their private concerns. All, with varying degrees of ambivalence, leave their families behind for the glory-humiliation dyad.

This personality type is not, of course, limited to literature. Real-life cases can be multiplied. One thinks, for example, of J. Robert Oppenheimer who, after World War II, began to be harassed by the authorities for alleged communist sympathies. The real reason behind the harassment was that Oppenheimer started speaking out in favor of international arms control and against the hydrogen bomb, which he saw as a genocidal weapon. The Air Force leadership and members of the Atomic Energy Commission, who were invested in nuclear buildup, became upset with him. Lewis Strauss, the head of the Atomic Energy Commission, decided to damage Oppenheimer's public standing by branding him a communist. The Commission gave Oppenheimer a choice to either give up his security clearance or ask for a hearing. Oppenheimer could have simply walked away. He could have ignored the attempts to

[30] Ibid., p. 126. [31] Ibid.

discredit him. Like Kohlhaas, he could have counted his blessings and cut his losses. After all, he was the director of the Institute for Advanced Studies in Princeton, he was the most famous scientist in America, he had a spouse and two children, and he could have shrugged off the government's threats. As his colleague at the Institute, Albert Einstein, reportedly said to him: "Robert, you are Mr. Atomic. You don't need them, they need you. Just walk away. Why should you go through this?"[32] But Oppenheimer did not listen. He could not stand the humiliation of being branded unsuitable to hold state secrets. When Einstein saw him leave for the hearings, he reportedly told his secretary – "[T]here goes a nar" (nar is Yiddish for a fool). Several weeks later, his name dragged through the mud in a series of rigged hearings, Mr. Atomic returned to Princeton, stripped of his clearance, a shell of a man, and remained aloof from politics for the rest of his life. In 2022, fifty-five years after he died, the Department of Energy reversed its decision to cancel Oppenheimer's clearance. For Oppenheimer and his family that was far too late.

Elizabeth Barrett Browning's poem, "Mother and Poet,"[33] offers a moving meditation on the private costs of living in and for great public causes. The speaker in that poem, a patriotic Italian mother who loses both of her sons in the Italian wars of unification, comes to realize that her public spiritedness has led to a terrible private agony. Unlike Hector, Ajax, and Michael Kohlhaas, the mother becomes disillusioned with her political commitments. She recalls, with a mix of pride and regret, how she inculcated love of country in her children:

> *I* taught them, no doubt,
> That a country's a thing men should die for at need.
> *I* prated of liberty, rights, and about
> The tyrant cast out.

But her patriotism has come to taste like ashes when its private costs have been tallied up. After she loses both of her sons she can't help but wonder what glory is for, if it is not for the living to enjoy:

> And, when Italy 's made, for what end is it done
> If we have not a son?

32 The episode is recounted in an interview with Kai Bird, coauthor of Oppenheimer's famous biography, in Christopher Cassel, *To End All War: Oppenheimer & the Atomic Bomb* (NBC/ Universal Documentary, 2023). For a fuller account, see the closing chapters of Kai Bird and Martin Sherwin, *American Prometheus: The Triumphs and Tragedy of J. Robert Oppenheimer* (New York: Vintage Books, 2006).

33 www.poetryfoundation.org/poems/43728/mother-and-poet.

The mother who starts out as an enthusiast for Italian liberty retreats from politics. The poem chillingly lays out how a commitment to grand purposes melts away before the reality of private costs. It is a song of regret for a life lived in these great causes, for the private price of being taken in by them. The political mother who raised her children to think it is good and honorable to die for Italy becomes estranged and alienated from Italy once her sons have actually died for it:

> Dead! One of them shot by the sea in the east,
> And one of them shot in the west by the sea.
> Both! both my boys! If in keeping the feast
> You want a great song for your Italy free,
> Let none look at *me* !

7.6 Summary

In this chapter, we took up the conceptual relationship between humiliation and glory. We argued that reversing humiliation has often been the Periclean cause for which glory is assigned. Leaning on James, Orwell, and Machiavelli, we explored the attraction of political causes in general and suggested that undoing a history of humiliation had a particular grip on the political imagination. We further noted that while glory is the reward for ending political humiliation, the sentiment itself is often understood in terms of lost glory. We proceeded to argue that the symmetry between the two terms was incomplete: While reversing political humiliation will always win glory, it is possible to become glorious for other causes, completely unrelated to that reversal. In Section 7.4 of the chapter, we noted that the humiliation/glory dyad has very different consequences for men and women: While war can be humiliating for both, glory is typically reserved for the men planning and fighting the wars. In so much as that is the case, war is essentially humiliating for women in a way it is not for men. We concluded by suggesting that a preoccupation with the humiliation/glory dynamic, a life lived in these "great causes," necessarily comes at the expense of the private, intimate milieu of the person who pursues them.

CHAPTER 8

Three Case Studies

In this chapter, I will offer three case studies to illustrate that the theory developed in these pages is helpful for understanding the entanglement of glory and humiliation in real-world cases. Each of the examples discussed illustrates the philosophical features of these dispositions, as well as the connections between them.

8.1 The Rise of ISIS

ISIS and its followers are preoccupied with (what they perceive as) the loss of Muslim honor; the group's propaganda materials tell a story of betrayal, humiliation, and subjugation of Muslims by Western powers. The organization's gruesome videos as well as key written materials are rife with references to such grievances – from the cynical treatment of the Arabs under the 1916 Sykes-Picot Agreement between Britain and France to the abuse of prisoners at the Abu Ghraib prison in Iraq by American guards in 2003 to the 2005 Haditha massacre in Iraq and the killings of Iraqi civilians by Blackwater contractors in Baghdad. A report by the Tony Blair Institute for Global Change argues that the need to reverse and counteract humiliation is a prominent theme in ISIS propaganda materials – more so than in materials distributed by competitors such as Al-Qaeda and Al-Nusra Front. Researchers found that 40 percent of ISIS propaganda references the need to end Muslim humiliation, compared to 31 percent of materials by Al-Qaeda in the Arabian Peninsula and 23 percent of Al-Nusra literature.[1]

ISIS propaganda materials typically tell a story of corrupt and weak Arab leaders who have allowed their people to be exploited by Westerners and, in the process, have emasculated themselves and the Arab national character. It is this process of emasculation that ISIS promises to reverse with its

[1] https://institute.global/sites/default/files/inlinefiles/IGC_Inside%20Jihadi%20Mind_18.08.17.pdf.

hyper-violent, perversely telegenic brand. A 2014 VICE News documentary shows fighters bulldozing the Sykes-Picot borderline between Iraq and Syria. As they stomp on the vacant ground where the fence separating the two countries once stood, the fighters shout: "[W]e have broken Sykes-Picot." The video also features interviews with militants who proudly proclaim the "reunification" of Arab lands and with civilians who applaud the ability to now travel between these territories without a passport.[2]

A popular "nasheed" or fighting song that was disseminated among potential ISIS recruits opens with the line, "A life of humiliation: no, I am not content with it."[3] Another propaganda video aimed at Syrian refugees features the now infamous footage of a journalist kicking a Syrian refugee escaping authorities in Hungary. The narrator highlights the indignity and calls for refugees to escape humiliation in Europe and come back to fight for the Caliphate.[4]

In the first issue of Dabiq, the organization's promotional magazine, ISIS spokesperson Abu Muhammed al-Adnani writes: "The time has come for those generations that were drowning in oceans of disgrace, being nursed on the milk of humiliation, and being ruled by the vilest of all people, after their long slumber in the darkness of neglect—the time has come for them to rise. The time has come for the ummah [Muslim community – NE] of Muhammad . . . to wake up from its sleep, remove the garments of dishonor, and shake off the dust of humiliation and disgrace, for the era of lamenting and moaning has gone, and the dawn of honor has emerged anew. The sun of jihad has risen. The glad tidings of good are shining. Triumph looms on the horizon. The signs of victory have appeared."[5]

A 2015 video released by the Information Office for the Islamic State chides Muslims who tried to escape from Syria to France, Germany, and other European countries. The video features an official who scolds would-be refugees and asks: "All those leaving our state for the lands of unbelief – shame on you! . . . What life do you hope [to live there]?! A life of humiliation and submission?! What life of ease are you entering?! And in the countries that are responsible for your destruction?!" A second speaker adds: "Those who emigrate to the lands of unbelief, especially France – we tell you: You will face problems in terms of wearing the niqab and the

[2] www.youtube.com/watch?v=TxX_THjtXOw.

[3] Quoted in full in the Blair Institute Report, https://institute.global/sites/default/files/inlinefiles/IGC_Inside%20Jihadi%20Mind_18.08.17.pdf, p. 26.
For a further discussion, see https://clarionproject.org/dominance-humiliation-rape/.

[4] www.thetimes.co.uk/article/come-join-the-caliphate-isis-urges-refugees-2z5ztm993qd.

[5] https://eaworldview.com/2014/06/iraq-text-isis-declaration-caliphate-muslims/.

hijab [traditional body and head covers worn by some Muslim women – NE]. You will face difficulties throughout your lives. You will not have honor in these countries, but only humiliation . . . You have no honor. Honor is only found in jihad and the land of the caliphate."[6]

The *Management of Savagery*, an important handbook distributed widely among ISIS recruits also describes a transition of Muslims from "domesticated" passivity and subjugation to militarized agency: "Those who have not boldly entered wars during their lifetimes do not understand the role of violence and coarseness against the infidels in combat and media battles. The stage of domesticating the Muslims which they have already passed through has had an effect on them. The reality of this role must be understood by explaining it to the youth who want to fight. They are different from the Arabs at the beginning of the Prophet's mission. The Arabs used to fight and know the nature of wars."[7]

Finally, in a famous speech given on July 1, 2014, immediately after the proclamation of the Caliphate, ISIS leader Abu Bakr al-Baghdadi has the following to say: "[Following the fall of the historical Caliphate the] disbelievers [sought to] humiliate the Muslims, dominate them in every region, plunder their wealth and resources, and rob them of their rights." The new Caliph then adds: "[T]hose rulers continue striving to enslave the Muslims, pulling them away from their religion with those slogans. So either the Muslim pulls away from his religion, disbelieves in Allah, and disgracefully submits to the man-made shirk (polytheistic) laws of the east and west, living despicably and disgracefully as a follower, by repeating those slogans without will and honor, or he lives persecuted, targeted, and expelled, to end up being killed, imprisoned, or terribly tortured, on the accusation of terrorism. Because terrorism is to disbelieve in those slogans and to believe in Allah. Terrorism is to refer to Allah's law for judgement. Terrorism is to worship Allah as He ordered you. Terrorism is to refuse humiliation, subjugation, and subordination [to the kuffār – infidels]. Terrorism is for the Muslim to live as a Muslim, honorably with might and freedom. Terrorism is to insist upon your rights and not give them up."[8]

[6] www.memri.org/jttm/isis-fighters-refugees-do-not-migrate-france-or-germany-migrate-immediately-islamic-state#_ednref1.

[7] The full text of the Management of Savagery is available here: https://universityofleeds.github.io/philtaylorpapers/pmt/exhibits/2800/Management_of_Savagery.pdf.
For an excellent discussion of the centrality of this text for ISIS, see Fawaz Gerges, *ISIS: A History* (Princeton, NJ: Princeton University Press, 2017): chapter 1.

[8] See Haroro Ingram, Craig Whiteside and Charlie Winter, *The ISIS Reader: Milestone Texts of the Islamic State Movement* (New York: Oxford University Press, 2020): chapter 7.

The centrality of humiliation in ISIS propaganda and ideology sheds light on a long-standing debate about the causes and sources of suicidal terrorism. The emphasis ISIS places on humiliation in its list of grievances lends credibility to the claims of writers like Robert Pape and Peter Bergen who have argued that a desire to counteract perceived humiliation is more important in accounting for suicidal terrorism than material distress or innate animosity towards the West. The most famous version of this argument can be found in Pape's 2005 book *Dying to Win.*[9] Echoing Pape, Peter Bergen and Michael Lind wrote, two years later, that "two factors link suicidal terrorists: first, they are members of communities that feel humiliated by genuine or perceived occupation . . . Second, suicide bombers seek to change the policies of democratic occupying powers like Israel and the United States by influencing their public opinion–in a sense making the occupying power suffer the same level of humiliation they have felt."[10] For Bergen, Lind, and Pape, this conclusion suggests a need to reduce and remove flash points that can cause humiliation – such as American bases in key Muslim locales and other traditional perceived humiliations like the Israeli control of important Muslim religious sites. According to this theory, removing these flash points would either reduce the occurrence of suicidal terrorism or, at least, undermine the narrative of grievance that helps to bring in new recruits.

While ISIS' use of humiliation as a recruitment tool lends some credence to the thesis these authors advance, there is something unusual about the organization's preoccupation with the term. ISIS was not, of course, the first to harness a sense of historical humiliation for political purposes. To rehash two key twentieth-century examples we have already mentioned – the Nazis masterfully played on the humiliation of Germany after World War I to enlist popular support; Nasser and later Sadat used the humiliation of Arabs in the 1967 war with Israel to spur their people towards the retaliatory war that eventually came in 1973. But there is an important difference. Whereas claims of humiliation are often used to galvanize a nation and stir people to shift their interest from the private to the national sphere, ISIS deployed these claims to *create* a nation. The political aim that ISIS was hoping to achieve through its propaganda of humiliation was not to buttress the motivation of fighters so that they may be willing to die for the Islamic State but, rather, to plant the idea of an

[9] Robert Pape, *Dying to Win: The Strategic Logic of Suicide Terrorism* (New York: Random House, 2005).

[10] https://democracyjournal.org/magazine/3/a-matter-of-pride/.

Islamic State in people who did not have it – in volunteers and recruits who came from divergent ethnic, racial, and geographical backgrounds and had little in common.

In his famous essay "What is a Nation," Ernest Renan claims that a nation consists of a shared legacy and a desire to perpetuate that legacy into the future. The legacy is rarely accurate, and it consists in a mix of remembering and forgetting. The key, for Renan, is to have a joint venture that participants want to perpetuate or project into the future: "A nation is a soul, a spiritual principle. Two things, which in truth are but one, constitute this soul or spiritual principle. One lies in the past, one in the present. One is the possession in common of a rich legacy of memories; the other is present-day consent, the desire to live together, the will to perpetuate the value of the heritage that one has received in an undivided form . . . The nation, like the individual, is the culmination of a long past of endeavors, sacrifice, and devotion . . . One loves in proportion to the sacrifices to which one has consented, and in proportion to the ills that one has suffered. One loves the house that one has built and that one has handed down. The Spartan song – "We are what you were; we will be what you are" – is, in its simplicity, the abridged hymn of every *patrie*."[11]

By focusing as it did on a history of past humiliations (some real, some imagined), ISIS manages to manufacture a key component in the formation of national identity – a legacy that disparate fighters can harken back to. When this is combined with the organization's fetishistic fixation on the revival of seventh-century Islam – from the design of its flag to the length of the trousers fighters wear to the reinstatement of sexual slavery and protection taxes for non-Muslims, the picture emerges of an organization that is manufacturing a past full of glories and pains – an instant, packaged history, that can, as it were, be handed down to its proponents.

But how does ISIS' preoccupation with political humiliation correspond with the theory offered in this book? The claims made by the organization illustrate each of the features of the account of humiliation in international relations discussed in Chapter 5: The Sykes-Picot Agreement and the French-British betrayal of the Arab forces who revolted against the Ottomans is offered up as a historical example that the Arabs could be done to at will, that their own national aspirations and projects, much like the Melians', didn't matter to the great powers. The Sykes-Picot border, a

[11] Ernest Renan, "What is a Nation?" Text of a Lecture Delivered at the Sorbonne on March 11th, 1882, in *Qu'est-ce qu'une nation?* trans. Rundell (Paris, France: Presses Pocket, 1992). http://ucparis.fr/files/9313/6549/9943/What_is_a_Nation.pdf.

straight line made in violation of promises to Sharif Hussein and other leaders, is taken by ISIS as a geographical marker of that humiliation, as a concrete reminder of the fact that the tribal realities and geographical contingencies of Arab lives mattered not at all to the British and French officials who divided up the area. The Arabs were not seen as subjects to whom promises must be kept. All that mattered was access to their oil and the ability to transport it to the Mediterranean.

ISIS' successful recruitment efforts also show how the sense of humiliation travels in time and space: As we noted, one of the organization's first videos featured the bulldozing of the Sykes-Picot border between Iraq and Syria, and the fighters featured in that clip referred prominently to that old humiliation – more than a century in the making – as the source of their rage. ISIS' success in recruiting young middle and upper-middle-class Muslim youths in Europe exemplifies that a sense of humiliation can transcend space as well as time – those youngsters were usually not the immediate recipients of the humiliations ISIS spoke of. In fact, by many measures of success, their lives were going quite well. And yet, when ISIS called, they went. That fact powerfully demonstrates the motility of the sense of humiliation, it reminds us that we can be humiliated, as it were, by proxy.

ISIS propaganda showcases another feature of humiliation that we have discussed – the independence of the sense of humiliation from the need for recognition. ISIS fighters report being humiliated by infidels and apostates, by members of groups that they do not view as full moral agents. ISIS leaders and ISIS jurisprudence do not recognize or seek recognition from Shiites, apostate Sunnis, Jews, or Christians. And yet all of these "enemies" are guilty of humiliating the organization's proponents.

Furthermore, ISIS pronouncements support the claim we made about the phenomenology of humiliation – that it consists, importantly, in the sense of being replaced and displaced. Muslims, so the ISIS propaganda line goes, had a history of being ascendant in both martial and cultural affairs. But in recent centuries, they have been displaced by "crusaders" and "apostates." They have been replaced by those who are, as far as the group's leaders are concerned, inferior to them. The organization's promise is one of restoration.

Finally, the story of ISIS stands out as a case study illuminating the relationship between historical humiliation and glory. The project of reversing the humiliation of Muslims, the aim of recreating the Caliphate is supposed to win glory for those who fight for ISIS; that glory was to be found in ending Muslims' disgrace and dishonor. As al-Adnani put it, "[T]he time has come for the ummah [Muslim community – NE] of Muhammad ... to wake up from its sleep, remove the garments of dishonor, and shake off the dust of humiliation and disgrace, for the era of lamenting and moaning has gone, and the dawn of honor has emerged anew. The sun of jihad has risen." The humiliation ISIS pointed to was defined as lost glory; reversing that humiliation, ending the sense of historical disgrace, was the political cause or purpose that would win glory for the organization's fighters.

8.2 Russia's Revanchist Foreign Policy

Russia's foreign policy in the last two decades provides another demonstration of the theory on offer here. More specifically, the sense of replacement is key to understanding the Kremlin's recent actions. From the annexation of Crimea to its involvement in Syria to the 2022 invasion of Ukraine, many of Russia's moves in the international arena have been framed as responses to being marginalized and dismissed after the collapse of the Soviet Union. Putin and his officials consistently claim that Russia has been relegated to second-rate power status after 1990, that it has been disrespected, discarded, and replaced on the world stage, and that they (and only they) can halt this decline. Some of the primary examples Russian leaders provide include the NATO intervention in Bosnia in 1995 against Russia's wishes, the quick dismissal of Russia's offer to help prosecute the "War on Terror" after the attacks of September 11 (specifically the offer to help with intelligence in Afghanistan), the beginning of the 2003 war in Iraq (again against Russia's wishes), and the second round of NATO expansion in 2004 which removed the former Soviet Republics of Estonia, Latvia, and Lithuania from Russia's traditional sphere of influence.

In a recent book, journalist Shaun Walker relays a phone conversation between Bill Clinton and Russia's Boris Yeltsin in which the former ignored the latter's objections to a planned bombing campaign in Serbia. A younger Vladimir Putin, who learned of the exchange, was shocked to discover how easy it had become to dismiss Russia's interests and vowed to change that.[12] Immediately before becoming president, Putin wrote the following in the newspaper Nezavisimaya Gazeta: "[F]or the first time in the past 200–300 years, Russia faces the real danger of being demoted to

[12] Walker, *The Long Hangover*. See also Shaun Walker, "The Humiliation that Pushed Putin to Try and Recapture Russian Glory," *History* (February 23, 2018). www.history.com/news/vladimir-putin-russia-power.

the second, or even the third tier of global powers. To prevent this from happening, we as a nation are to pull all our intellectual, physical, and normative resources, thus ensuring that ours remains a first-tier nation. Our efforts must be united, coherent and constructive, for no one else would do this work for us."[13] Shortly after that, in an "Open Letter to Voters" published in the *Izvestia, Kommersant,* and *Komsomolskaya Pravda* newspapers on February 25, 2000, Putin added: "Our priority is to restore the personal dignity of the people in the name of the dignity of the nation ... Russia has long ceased to be just a reduced map of the Soviet Union; it is a confident power with a great future and a great people ... The past decade has brought dramatic changes to the consciousness of the people. Our citizens are not yet rich, but they are independent and self-confident ... Our army is emerging from a prolonged crisis with honor and is becoming ever more efficient and professional ... True, Russia has ceased to be an empire, but it has not wasted its potential as a great power ... It is unreasonable to be afraid of a strong Russia, but it should be reckoned with. Insulting us is counterproductive."[14]

The anger about the diminution of Russia, the resentment about its displacement as a great power, and the desire to see it reinstated come across in interviews with some leading Russian intellectuals and journalists as well. Here is director and screenwriter Andrei Konchalovsky in an interview with *Vesti Nedeli* on January 24, 2009: "Clinton was good to Russia, because Russia did not exist on the map. Clinton was good to Russia because Russia had kneeled before America for it was living on American money for a decade. Today's Russia is no longer kneeling and has the means to defend its interests on the global stage. No one likes it, but they have to deal with it. I do not think that Russia and the US can be friends. Nevertheless, their relationship can be mutually respectful. Russia's interests must be acknowledged and accepted as given, as the status quo."[15] And, more to the point, the journalist Vladimir Pozner had the following to say in an interview with the newspaper *Argumenty i Fakty* on April 2, 2015: "[Imperial ambitions] are like a phantom-limb pain. When western states tell us, 'you are not great ... you are second- or even fifth-tier nation [and] your time has passed ...' it only adds to those conscious and subconscious grievances over the lost empire. Yet, people

[13] Vladimir Putin, "Russia at the Turn of the Millennium," *Nezavisimaya Gazeta* (December 30, 1999) (Russian translation courtesy of Alan Zebek).

[14] http://en.kremlin.ru/events/president/transcripts/24144.

[15] www.vesti.ru/article/2162310 (Russian translation courtesy of Alan Zebek).

still deem themselves as part of a great power. Telling them that it is no longer the case, only exacerbates their sense of loss."[16]

The sense of humiliation understood as lost ascendency is palpable in these excerpts. So is the sense of being replaced, so central to the phenomenology of humiliation – in this case, Russia being displaced from a position of leadership on the world stage and being replaced by other powers. These ideas of historical humiliation and the need to reverse it have served Putin for years as a means of framing his foreign policy. Their significance grew exponentially as Putin and his officials started explaining their aggressive war in Ukraine to their own people and to the rest of the world.

In his February 21, 2022 speech,[17] given a week before the invasion, Putin described contemporary Ukraine as Lenin's creation. The Soviet leader was the "author and architect" of that state but now, Putin complained, the "ungrateful" Ukrainians were destroying statues of Lenin as part of their de-communizing process. If the Ukrainians wanted to rid themselves of communism's legacy, Putin added, menacingly, Russia would help them do that thoroughly, by taking back all the territories the Bolsheviks originally gave them. And speaking of ingratitude, Putin also complained that Ukrainians were removing statues of Russian heroes such as the eighteenth-century General Alexander Suvorov without whom Ukraine would not have many of its current assets.

If Ukraine became an independent nation, it was not so much because of its own push for nationhood, Putin continued, but rather it was due to mistakes the Soviet Union's leadership made. After the breakup of the Soviet Union, Russia continued to treat Ukraine respectfully – as evidenced by the preferred trade partner status it granted the Ukrainians – only to be rewarded with brashness, and with Ukraine dangling the possibility of getting closer with Russia as a way to blackmail the West for benefits. Contemporary Ukraine was persecuting ethnic Russians in its territories and was fast becoming a staging ground for Western powers who wanted to harm Russia. The promises of the 1990s according to which German reunification would not result in any NATO expansion eastward turned out to be empty. NATO and American presence in Ukraine constituted a "knife to the throat" for Russia – because American weapons and radar systems now covered all of its territory and threatened its independence. When Russian concerns about this were

[16] https://aif.ru/politics/russia/1438503 (Russian translation courtesy of Alan Zebek).
[17] http://en.kremlin.ru/events/president/news/67828.

voiced they were brushed off. NATO "completely ignor[ed][Russian] concerns protests and warnings. Unfortunately, they just spit on them and did whatever they wanted."

Three days later, Putin followed up with another speech that rehashed many of the same grievances:[18] The decline of Soviet power led to "complete degradation and oblivion." The Russian state that emerged in the 1990s was weak, and the weak are easily humiliated. After the collapse of the Soviet Union, the United States tried to lull Russia into a sense of false security with promises that NATO would not expand. At the same time, the Americans tried to weaken traditional Russian values by supporting all varieties of subversive elements inside the country. Russia tried, in good faith, to negotiate limits on NATO expansion as well as some of its other interests with the Americans, but without success. Russia sees NATO and Western presence so close to its borders as an existential threat and wants to avoid making the mistake of waiting too long before countering the threat. The "special military operation" in Ukraine was meant to save ethno-Russians in Donetsk and Luhansk from the humiliation of Ukraine dismissing their cultural and national heritage and to protect them from being ethnically cleansed by the Ukrainians. Russia, in other words, having been a victim of humiliation by the West, would help those who depend on it to avoid similar humiliations.

In a third speech,[19] given on March 16, a little more than two weeks into the war, Putin reiterated that Western countries have made it their aim since the 1990s to turn Russia into a weak and defanged country. Perhaps other nations are used to being humiliated and living by Western dictates, he suggested, but Russia will not suffer that fate; it won't be forced to surrender its security interests or its prominence due to military threats or the imposition of sanctions.

Putin has also been explicit in arguing that his "special military operation" in Ukraine will be a source of great glory for Russia. At a speech marking eight years of the annexation of Crimea, he noted that the invasion started on the birthday of Fyodor Ushakov, a famous eighteenth-century Russian naval commander, whom Putin quotes as saying that "these storms [supposedly battles during Ushakov's lifetime] will contribute to Russia's glory."[20] Putin's Victory Day speech in 2022 (the celebration of Russia's victory in World War II) ended on a similar note: There was glory in fighting a war of self-defense against NATO – in

[18] http://en.kremlin.ru/events/president/news/67843.
[19] www.youtube.com/watch?v=7FyFkAyqn4Q. [20] www.informationliberation.com/?id=62965.

supporting an effort crucial for the preservation of the country's identity and for reinstating its honor.[21]

Finally, in an event marking the 350th birthday of Peter the Great, Putin spoke to a group of young entrepreneurs and celebrated the Tsar's boldness in expanding Russia's territory and influence during the so-called "Northern Wars." He also credited him for creating a new capital in St. Petersburg. Dismissing the idea that Peter was an occupier, Putin described him as "returning" and "strengthening" Slavic territories to Russia. Then, he added, in reference to the war in Ukraine: "[W]ell, apparently, it has also fallen to us to return and to strengthen."[22]

I am not focused here on the historical accuracy of Putin's statements. The version of history he presents is selective, manipulative, and quite often simply false. What I am interested in here is the role political humiliation plays in the lead up to war, in the way leaders use (and abuse) a narrative of historical humiliation to call recruits to the flag. The sense of betrayal, bitterness, and national humiliation in Putin's addresses is difficult to miss. And one readily spots in them the key elements of the theory of political humiliation we have outlined: After the fall of the Soviet Union, Russia was "done to" by the Western powers. It was no longer taken seriously as an actor whose views needed to be considered – much like the Melians were brushed off by the Athenians. Russia, according to Putin, was no longer taken to be a party to whom promises had to be kept; American promises to avoid NATO expansion were simply ignored because Russia was not strong enough to demand compliance (much like the Arab tribes were not strong enough to enforce British and French promises in the aftermath of World War I). Furthermore, as far as Putin was concerned, Russia's history and its sense of itself, its understanding of its own cultural and geopolitical identity, were ignored as NATO expanded into Russia's traditional sphere of influence and as Western powers encouraged liberalization and democratization in Ukraine and in the Baltics.

Being done to, suffering from broken promises, having one's heritage and sense of national identity ignored – all key elements of political humiliation enumerated in this book – are on full display in Putin's understanding of the world and in his justification for going to war in Ukraine. To reiterate: Putin's interpretation of history is unreliable and self-serving. But that is not our focus. Rather, I am noting the fact that

[21] www.bbc.com/news/world-europe-61377886.
[22] www.nytimes.com/2022/06/09/world/europe/putin-peter-the-great.html.

framing a conflict as a way of addressing historical humiliations, which is the strategy Putin chose, has remarkable political and psychological force. The story Russian officials have broadcast is this: Russia has been humiliated since the collapse of the Soviet Union. The invasion of Ukraine was part of the reversal of that humiliation. Putin would make Russia matter again; he would return it to the world stage, he would make others take its interests seriously again, he would, like Peter the Great, become a defender of Slavdom and Russian identity. In short, by reversing a history of humiliations, he would win himself and Russia glory.

The desire to humiliate Ukraine, so evident in Putin's speeches, the callous throwing around of Russia's weight (see, for example, his statement that if Ukraine wanted to decommunize, Russia would help it "decommunize all the way" by taking away territory handed over during Soviet times), the claim that Ukraine has no real history or identity as a country – are framed as a retaliation for the humiliation suffered by Russia. The psychological dynamics here are striking: Humiliation is purged by humiliating; those who suffered it redeem themselves by inflicting it. Hitler made the French sign their surrender in 1942 in the same train wagon that the Germans surrendered in at the end of the Great War, ISIS would portray its fighters as the executioners of "infidels," no longer their victims. Russia would stomp on the Western-leaning Ukraine as it says it was stomped on by the Americans and the West.

8.3 The 1973 Middle East War

On June 5, 1967, after months of mounting tensions, Israel launched a preemptive war against Egypt, Jordan, and Syria. In what turned out to be the decisive move of that war, the IDF destroyed the air forces of all three countries in the first hours of fighting. This gave the Israelis air superiority in all major theaters of the conflict and paved the way to a resounding victory. In six days, Israel captured the Sinai Peninsula and Gaza from Egypt, Eastern Jerusalem and the territory known today as the "West Bank" from Jordan, and the Golan Heights from Syria. These territorial gains more than tripled the areas that the Jewish state controlled.[23]

The war had profound geopolitical and psychological effects on everyone who fought in it. Israel had, almost overnight, turned from a young, anxious, insecure state, into a regional superpower. The swift victory

[23] For a solid account of the war's trajectory, see Michael Oren, *Six Days of War: June 1967 and the Making of the Modern Middle East* (New York: Presidio, 2003).

emboldened messianic elements in Israeli society, who interpreted the campaign's success as a sign of providential blessing and as a religious mandate for the state to expand further. The Israeli settlement movement was born, and it successfully lobbied for the creation of Jewish communities in the newly occupied territories (primarily in the West Bank).[24] The overwhelming military victory also generated a set of dismissive, arrogant attitudes among Israelis towards their Arab neighbors. It was not unusual to encounter songs and skits mocking the fighting abilities and commitment of Arab soldiers. Perhaps most indicative of these attitudes was the new literary genre of Six-Day War "victory albums," celebratory photobooks, which sold well, and provided pictorial evidence of the war's achievements. These albums regularly contained images of the shoes left behind by retreating Egyptian soldiers as well as captions dismissive of their military prowess. One such caption read: "In this war, Egyptian soldiers won in ... long distance running. They left behind thousands of pairs of shoes." Another caption that accompanied pictures of piles of shoes left behind by Egyptian soldiers in the Sinai Desert read: "They are not used to shoes: In many places in the Sinai concentrations of shoes were found, an indication of fleeing. The Egyptian soldier is not accustomed to wearing shoes, while his commanders aim to instruct him in launching rockets."[25] Other popular books collected jokes that displayed dismissive attitudes towards the state's defeated foes. One such book, titled *All the Nasser Jokes* (referring to the name of the Egyptian president at the time of the 1967 war), featured the following anecdote: A military instructor asks a trainee, "What steps will you take if you encounter an armed Israeli soldier?" "Big [steps] and quick as possible," the soldier answers.[26] Ezer Weizman, the commander of the Israeli Air Force until 1966 and the head of the IDF's Operations Sector during the war (and later, the state's minister of defense and its seventh president) told a reporter: "The Arabs? War is not for them. The stories about their military prowess are taken from the legends of Lawrence [of Arabia NE] and they are only legends."[27]

If the war instilled an arrogant, almost euphoric sense of confidence among many Israelis, its impact on the Arab world was starkly different.

[24] For an outstanding discussion of the war's aftermath in Israel, see Gershom Gorenberg, *The Accidental Empire: Israel and the Birth of the Settlements, 1967–1977* (New York: Holt, 2007).

[25] Alon Gan, "'Till Death,' 'Brothers Sitting in the Shadow of Cannons': Oz and Levin as Anti-Euphoric Prophets," in Ofer Shif and Aviva Halamish (eds.), *Israel 67–77 Continuity and Turning Point* (Beer Sheva, Israel: Ben-Gurion Institute, 2017): p. 338 (Hebrew, my translation).

[26] Ibid. (Hebrew, my translation).

[27] Ibid. (Hebrew, my translation).

The defeat generated widespread feelings of shame and humiliation among political leaders and ordinary citizens alike. This was particularly true in Egypt, which until then perceived itself as the flag bearer of Arab nationalism and as the region's de facto leader. More than forty years after the war, Mohamed el-Sayed Said, a political analyst and the deputy head of Egypt's Al-Ahram Center for Political and Strategic Studies, told the Voice of America that "the 1967 defeat was like throwing us all into the gates of Hell ... It shocked us to the bones, and obviously changed dramatically the chemistry of political and cultural life in the country."[28] In fact, so great was the shock and disbelief at the defeat, that even the name given to the campaign reflected a degree of denial – it was dubbed *al-naksa* or *the setback*. Some Arab intellectuals tried to blame the defeat not on the lackluster performance of their armies or the mistakes of their leaders but on a Western conspiracy to help Israel.[29] Describing the national trauma that the defeat inflicted on his people, President Nasser noted that "after this great catastrophe we were like a man who went out to the street to be hit by a tram or a car and lay both motionless and senseless on the ground."[30]

The sense of despair brought on by the war seriously damaged the hopes of Arab nationalism represented by Nasser and pervaded Egyptian popular and intellectual culture.[31] This national trauma was very much on Anwar Sadat's mind when he replaced Nasser as president in 1970. And it was not only Sadat and the Egyptian leadership that wanted to undo the humiliation of 1967 – the sentiment was a genuinely popular one. The years following the 1967 defeat saw regular demonstrations and disturbances in Egypt, perhaps most famously in Tahrir Square in Cairo, with crowds demanding retaliation against the Israelis in order to restore Egypt's honor. Soldiers posted on the Egyptian side of the Suez Canal reported their frustration and humiliation while observing the Israelis occupy their land. Thus, for example, Mahmoud Khalaf of the Egyptian 129th Commando

[28] www.voanews.com/a/a-13-2007-06-06-voa52-66779027/564941.html.

[29] Hisham Melhem, "The Arab World Has Never Recovered from the Loss of 1967," *Foreign Policy* (June 5, 2017). https://foreignpolicy.com/2017/06/05/the-arab-world-has-never-recovered-from-the-loss-of-1967/.

[30] Quoted in Ibrahim Zabad, "Egypt: Defeat and the Transformation of State and Society," in *The Aftermath of Defeats in War* (London, UK: Palgrave Macmillan, 2019): p. 48.

[31] For a good discussion of the war's social and intellectual ramifications in Egypt, see Fouad Ajami, *The Arab Predicament: Arab Political Thought and Practice since 1967* (Cambridge, UK: Cambridge University Press, 1992): parts 1, 2. For a look at the defeat's impact on popular culture, see Tamara Maatouk, "On the Heels of 1967: Chahine, Cinema, and Emotional Response(s) to the Defeat," *International Journal of Middle East Studies* 55(1) (2023): 25–42.

Regiment recounted for an Al Jazeera documentary: "[T]he Israeli flag . . . I cannot describe the feeling. My generation knows this feeling very well. Stretching for 167 kilometers the Israeli flag is up high on every kilometer, right before your eyes all along the canal. I swear, it nearly turned the color of the Suez Canal black."[32]

On October 6, 1973, Egypt and Syria launched a surprise attack against Israel on two fronts. The first week of the Yom Kippur War (the assault was launched on the Jewish Holy Day. The conflict is referred to in Egypt as the October War or the Ramadan War) was disastrous for Israel. Its defensive lines along the Suez Canal collapsed, several Egyptian divisions crossed the Canal into the Israeli-held side, the Syrians made significant gains in the occupied Golan Heights, and the IDF lost about a third of its air force and a third of its tanks. Israel's military and political leaders, still basking in the glow of the 1967 campaign, were stunned, and for a few days believed that the state's existence was at stake. Some military leaders suggested that the country had to use its nuclear weapons as a last resort – an idea dismissed by Prime Minister Golda Meir.[33] But by the end of the war's second week, the tables had turned. The Syrians were repelled from the Golan Heights and the Israeli counteroffensive pushed deep into Syrian territory. In the Sinai, the Israelis managed, with great difficulty, to counterattack, crossed the Suez Canal into the Egyptian side, and by October 20, had moved armored divisions to the Egyptian side of the canal. By the time a ceasefire was declared, the IDF had encircled Egypt's Third Army and was in a position to destroy it. Israel had ultimately managed to turn the tide of the war but at a tremendous cost. During the eighteen days of fighting, it suffered 2,297 killed and 7,251 injured. Egypt and Syria also captured 314 Israeli prisoners of war.[34] These were staggering numbers considering the size of the population (approximately three million at the time). The sense of invincibility that set in after 1967 imploded, and the country went into a prolonged period of self-examination.[35]

[32] *The War in October* (Al Jazeera Documentary, 2013): part 1, 5 minutes and 40 seconds in.

[33] See T. V. Paul, "Taboo or Tradition? The Non-Use of Nuclear Weapons in World Politics," *Review of International Studies* 36(4) (2010): 853–63, 859.

[34] Uri Bar-Joseph, "Lessons not Learned: Israel in the Post-Yom Kippur War Era," *Israel Affairs* 14(1) (2008): 70–83, 70.

[35] For an excellent general account of the war, see Abraham Rabinovich, *The Yom Kippur War: The Epic Encounter that Transformed the Middle East* (New York: Schocken Books, 2017). See also the more recent Uri Kaufman, *Eighteen Days in October: The Yom Kippur War and How It Created the Modern Middle East* (New York: St Martin's Press, 2023).

The 1973 war presents an especially interesting case study for examining the dynamics of humiliation and glory that we have been discussing. First, the war was launched to address a historical humiliation: Sadat's main purpose in attacking Israel was not to destroy it or regain the territories the Israelis had captured from Egypt in 1967. He did not think those were realistic goals. Rather, he wanted to make limited territorial gains and undo the humiliation the Arab armies had suffered in the previous engagement. Doing so, he thought, would change the perception of Arab helplessness and Israeli invincibility that resulted from the 1967 defeat.[36] Reflecting on the Egyptian leader's motives, the historian David Tal writes: "Sadat and his senior military commanders knew full well that the Egyptian army would not be able to repel the Israeli forces from all of the Sinai ... Sadat relates how Nasser told him that 'if we get even 10 cm of the Sinai, and [are able] to entrench there in a way that no power could remove us from there ... we will remove the shame that we carried from the 1967 defeat.' And indeed, this was precisely Sadat's war plan. He aimed for a symbolic yet significant achievement, occupying a narrow strip of land on the eastern banks of the Suez Canal ... Sadat did not intend to send his military to capture the whole of the Sinai, and the war need not result in Sinai's reoccupation. The goal of the attack would be to occupy that small piece of land and 'to entrench there in a way that no power could remove us from there.' This would be enough to declare victory and to announce that the shame of the humiliation had been removed."[37]

In his biography, Sadat repeatedly described the humiliation he personally felt after the 1967 defeat and his determination to address it. Thus, for example, he recounts his thoughts as the 1967 war drew to a close: "On

[36] Kissinger did not initially believe that Sadat was willing to start a war primarily to erase a sense of national disgrace, and that for him, erasing that sense of disgrace was a precondition to any diplomatic progress. Kissinger writes in his memoirs: "Every Israeli (and American) analysis before October 1973 agreed that Egypt and Syria lacked the capability to regain their territory by force of arms; hence it was assumed they would not attack. The premises were correct. The conclusions were not. What literally no one understood beforehand was the mind of Anwar Sadat. In his view, serious diplomacy was impossible while Israel considered itself militarily supreme and Egypt was paralyzed by humiliation. In 1972 he expelled Soviet troops from his country because of the disrespect shown by Soviet leaders toward Egyptians but above all because they would surely seek to impede his planned military move or else exploit it for Soviet ends. The following year he fought a war not to acquire a specific sliver of territory but to restore Egypt's self-respect and thereby increase its diplomatic flexibility. Clearly, there had been an intelligence failure. What no one believed—the consumers no more than the producers of intelligence—was the notion of starting an unwinnable war to restore self-respect." See Henry Kissinger, *Years of Upheaval* (Boston, MA: Little, Brown, 1982): p. 459.

[37] David Tal, "Who Needed the October 1973 War?" *Middle Eastern Studies* 52(5) (2016): 748–49.

June 9, still bewildered, I listened to a communique on the radio, issued by the general command, which said that the Israelis had crossed to the West Bank of the Canal and appealed to the world to judge this move. The communique implied such disgraceful helplessness, submissiveness, and humiliation it made my blood boil. I rose at once, put on my 'resistance' overalls, took my rifle with the telescopic lens, got into a small Fiat I had borrowed from Intelligence, and drove away to fight my own battle. It seemed much more honorable to die on the battlefield than to stay idle at home."[38]

The stark impressions of defeat did not fade with time. Describing his mood in 1970, Sadat writes: "I myself was completely overwhelmed by our defeat. It sank into the very fabric of my consciousness so that I relived it day and night. As its real dimensions were daily revealed to me, my agony intensified— and my sense of helplessness. I shut myself up for three whole weeks at my home near the Pyramids, concentrating on the recent events and trying, with all the fortitude I possessed, to weather the fierce campaign of denigration launched by both friend and foe. Doubts were cast on our fighting ability. It was alleged that the Egyptians were not fit for fighting, that no further wars could ever be fought for the recovery of our land and the restoration of our honor. If it were true, I thought, it could mean death and destruction for our people for thousands of years to come ... What a disgrace and humiliation ... I was brought up to love and believe in Egypt, and to have infinite faith in the Egyptian man. Would all this now go overboard? If it did, I would too. For in that case, I could never recognize the man I call myself, never know who I really was, but must live without identity—a stranger among strangers. Would such a life be worth living?"[39]

Thinking back on this fundamental motivation to attack Israel in 1973, Sadat writes: "The basic task was to wipe out the disgrace and humiliation that followed from the 1967 defeat. I reckoned it would be 1,000 times more honorable for us — 40,000 of my sons in the armed forces and myself — to be buried crossing the Canal than to accept such disgrace. Posterity would say we had died honorably on the battlefield ... and posterity would carry on the struggle."[40]

And, here he is again, affirming that the reversal of humiliation was the key strategic aim of the war; that it was not about winning back territory but about winning back the country's dignity: "In drawing up the general

[38] Anwar Sadat, *In Search of Identity: An Autobiography* (New York: Harper & Row, 1982): p. 177.
[39] Ibid., p. 184. [40] Ibid., p. 215.

strategic plan for the battle, I took many things into account, but primarily the strategic basis from which we ought to proceed. I used to tell Nasser that if we could recapture even 4 inches of Sinai territory (by which I meant a foothold, pure and simple), and establish ourselves there so firmly that no power on earth could dislodge us, then the whole situation would change — east, west, all over. First to go would be the humiliation we had endured since the 1967 defeat; for, to cross into Sinai and hold on to any territory recaptured would restore our self-confidence."[41]

If the weight of humiliation in 1967 moved Sadat to start a war in 1973, it also guided American Secretary of State Henry Kissinger as he worked to end it. Kissinger and President Nixon authorized a massive American resupply mission that replenished Israeli stockpiles and helped the IDF eventually turn the tide of the conflict. But the secretary of state also understood very clearly that if there was a chance of negotiating a future peace agreement in the Middle East, the war had to end with neither of the combatants completely humiliated. Two days after a ceasefire was reached, the IDF surrounded Egypt's Third Army and refused to let essential supplies through. Had Israel destroyed this massive force, Egypt's initial achievements in the war would have dissipated. Kissinger, who by now understood that Sadat started the war in great part to redeem his country's self-respect, was adamant about preventing this eventuality. He pressured the Israelis to allow supplies to reach the besieged Egyptian soldiers. But the Israelis stalled. Finally, an exasperated Kissinger berated the Israeli ambassador to Washington, Simcha Dinitz: "I have to say again your course is suicidal. You will not be permitted to destroy this army. You are destroying the possibility for negotiations."[42] Kissinger, then, enforced a kind of tie on the combatants, allowing each to save face. The United States was not willing to allow the Israelis to turn things around so decisively as to risk a repeat of the 1967 humiliation. As Wat Cluverius, a state department desk officer during the war told historian Kenneth Stein in an interview: "[W]e quickly understood that what we needed was that there would be no unchallenged winners and no humiliated losers. All of us agreed about that."[43] Stein continues to unequivocally describe the point of Kissinger's decision to rescue the Egyptian Third Army: "Washington's intervention saved one of Sadat's defeated armies from

[41] Ibid., p. 244. [42] Kissinger, *Years of Upheaval*, p. 609.

[43] Oded Rabinovich and Kenneth Stein, "The October 1973 War: Superpower Engagement and Estrangement," (Hebrew) *Zmanim* 84 (2003): 59–69, 65 (my translation).

decimation. The Americans thus thwarted Egypt 's comprehensive defeat, and allowed the Israelis only a modest victory in the October War."[44]

To recap, humiliation was a key factor in Egypt's decision to go to war. The October War was embraced on all levels of Egyptian society because of the possibility of redemption it offered. The Syrians, led by Hafez al-Assad, who was a senior military and defense official during the 1967 war, also felt the sting of defeat keenly and were motivated in part by similar reasons.[45] If the war was launched to reverse a humiliation, the sentiment also played a key role in bringing it to a close. Kissinger, in particular, whose shuttle diplomacy is rightly credited for finally bringing about a stable ceasefire, was convinced he had to prevent another Arab humiliation if his efforts at stopping the war and laying the groundwork for future peace were to stand a chance. In his mediation efforts he actively "engineered" that outcome.[46]

Humiliation was not only a consideration in starting and ending the 1973 war. It also featured prominently during the conflict as a *way* of war. As we discussed in Chapter 5, humiliation of one's enemies has historically served as an important military tactic. In the first days of the war, with the element of surprise on its side, Egypt took a large number of Israeli prisoners of war. Egyptian officials made sure that the demoralized and very obviously shaken Israeli captives, with their dusty, torn, and disheveled uniforms, were shown on Egyptian media. These images did much to boost Egyptian morale and to make the case that the humiliation of 1967 had been erased. Sameh Seif Elyazal, an Egyptian military intelligence officer, interviewed for an Al Jazeera documentary about the war, told the filmmaker that "the arrival of the first prisoners lifted the morale of all our forces. Straightaway Egyptian military intelligence resorted to photographing them and putting those photos in the media to try to boost the morale not only of the armed forces but also of the Egyptian people."[47]

[44] Ibid., 66 (my translation).

[45] On this point, see, for example, Sterett Pope, "The Education of Hafez al-Assad," *Worldview* 28(4) (1985): 4–8, and Reuben Miller, "The Israeli-Syrian Negotiations," *Mediterranean Quarterly* 11(4) (2000): 117–39. Miller writes: "[T]he 1967 war left a humiliating and indelible impression on Assad who served as Chief of Syrian Airforce and Defense Minister" (129).

[46] It's worth noting that Kissinger was also keen to avoid American humiliation in the war. He was very clear that the United States could not let two regional powers (Syria and Egypt) who were supplied by Soviet weapons defeat a regional actor (Israel) supplied by American arms. So just as Kissinger was intent to prevent the Israelis from humiliating the Egyptians, he was also careful not to allow the Soviets to humiliate the Americans in the larger proxy contest. See, generally, Kissinger, *Years of Upheaval*, and Stein, October 1973 War.

[47] *The War in October* (Al Jazeera): part 3, 28 minutes and 20 seconds in. www.youtube.com/watch?v=7fiC1UXZPGg.

After all, the images were the polar opposite of the iconic 1967 pictures that showed shell-shocked, barefooted Egyptian soldiers surrendering to the Israelis or, in some cases, running from them. The episode and its impact on Egyptians provides one more example of the psychological operation of humiliation, of the cruel symmetry at its heart: Those who are humiliated are often moved to inflict the exact same injury in return. Reflecting on this episode, as well as on the Compiègne Wagon incident we mentioned earlier (the Nazis insisting on the French surrendering to them in the same train car the Germans had surrendered to the French in at the end of World War I), and on other similar historical examples, it is hard to escape the conclusion that humiliation as a way of war, humiliating as a tactic, is deeply tied to the age-old dynamics of tit for tat revenge.

The Israelis, for their part, even after an American-Soviet sponsored ceasefire had been brokered on October 22, insisted on completing their encirclement of Egypt's Third Army on the western side of the Canal, cutting off approximately 30,000 men from food and medical supplies. As Kissinger notes in his memoirs, the Israelis dragged their feet in response to American demands to resupply the force and, for a while, seemed intent on tightening their siege, humiliating both the trapped soldiers and the Egyptian high command who had inflicted such a significant blow on the IDF in the first days of the war.[48] The humiliated Israelis could not resist inflicting humiliation on the helpless Third Army. But they eventually yielded to American and Soviet pressure and allowed the resupply of the besieged Egyptians in return for the release of the IDF prisoners. The termination of one humiliation was thus made conditional on the cessation of another.

The 1973 war illustrates another important feature of political humiliation that we have considered, namely that the feeling arises, with great intensity, quite apart from whether there are solid normative grounds for it to come about. As we have seen, there was a ubiquitous sense of national humiliation in Egypt after 1967. The desire to address that feeling was central in the run-up to the surprise attack on Israel. But, interestingly enough, that feeling was not morally grounded – at least not if we consult the account of humiliation in international affairs we laid out in Chapter 5. Israel did not launch its preemptive war in 1967 to humiliate its neighbors. It was, rather, genuinely worried about its survival at the time. The Egyptian (and Syrian) sense of humiliation did not spring from the fact

[48] For Israel's ignoring of American requests to resupply the Third Army, see Kissinger, *Years of Upheaval*, pp. 603–11.

that the Israelis had applied arbitrary force, had broken promises to them, disregarded their culture, or viewed them as depositories for natural resources. In other words, the Egyptian (and Syrian) sense of humiliation after 1967 did not spring primarily from a violation the Israelis had inflicted. Rather, they were humiliated because they lost a war that they assumed they would win, and they lost it quickly and resoundingly. Egypt, in particular, viewed itself as the leader of the Arab world in the late 1960s and as the vanguard of Arab nationalism, and the defeat in 1967 destroyed those perceptions. To win a war is not necessarily to humiliate the loser. To be vanquished does necessarily furnish normative grounds for feeling humiliated. Conceptually, the Egyptian response is closer to our philosophical understanding of embarrassment (being shown up as someone altogether different from what one thinks of oneself or tries to project to the world) than it is to our account of political humiliation.[49] And yet, as we noted earlier, such conceptual distinctions are neater than our everyday emotional experiences. Even though the acts of trying to humiliate, embarrass, or shame someone else are relatively easy to distinguish from each other, the feelings of being humiliated, embarrassed, or ashamed are more difficult to tease apart. Yes, Israel's response to its victory in 1967 (a victory which, it must be said, shocked the Israelis as much as it did anyone else) evinced a willingness to humiliate: The victory albums, photos, songs, and jokes at the expense of Egyptian soldiers were degrading. But these depictions of the Egyptians (it was primarily the Egyptians they were aimed at) were not the source of Egyptian feelings of humiliation. It's safe to say that most Egyptians did not read Israeli victory albums or listen to jokes made about them in Hebrew. The source of the humiliation was the defeat itself. The soul-searching and national crisis that Egypt sank into in the late 1960s emerged from the gap between the sense of national pride and exuberance that accompanied Nasser's version of pan-Arabism and the reality of comprehensive defeat. This historical episode illustrates for us, once more, that the most consequential and deeply felt humiliations don't necessarily follow any normative theory of humiliation. The Israelis did not humiliate the Egyptians in 1967. Israel did not do and Egypt did not suffer anything that is clearly humiliating under the normative theory of humiliation sketched in Chapter 5. And yet the Egyptians felt humiliated – humiliated enough to start a war that ultimately restored their self-respect, shook Israel to its core, almost started a nuclear war, and paved the way for a historic peace agreement in the late

49 See the discussion of the distinction between humiliation, shame, and embarrassment in Chapter 4.

1970s. None of this makes the normative account of humiliation offered earlier wrong. It just suggests that normative considerations are not the most important ones in understanding the dynamics of political humiliation.

The Yom Kippur War highlights yet another feature of political humiliation that we have discussed – its capacity to "travel." The sting of humiliation often impacts and mobilizes even those who are not directly impacted, sometimes those who live far from the locus of injury – as long as they feel some kind of affiliation with or affinity to the humiliated party. The surprise attack on Israel generated a great deal of excitement in the Arab world. Morocco ran large-scale blood drives to benefit the Egyptian and Syrian soldiers. Algeria gave $200 million to the Soviets in order to finance weapons and ammunition for the Egyptian and Syrian war effort. Iraq, Saudi Arabia, Jordan, and Morocco all sent troops to the Syrian front. In fact, the arrival of Iraqi forces to Syria slowed down the successful Israeli counteroffensive against the Syrians.[50] The fact that these governments contributed sometimes quite significant forces and resources to undo a humiliation that was primarily inflicted on others (Iraq, Morocco, and Saudi Arabia lost no territory in 1967), as well as the popular support for these policies, indicates that many in these countries felt that they shared in the 1967 humiliation, that, to use Margalit's term, the military disaster impacted their understanding of their "encompassing group." As in the case of ISIS' success in recruiting youths from Western countries, the 1973 case reminds us that it is possible to be humiliated by proxy.

Finally, the 1973 war is a reminder of the close ties between reversing humiliation and the bestowal of glory. The relative success of the Egyptian forces against Israel, particularly in the first days of the war, was seen by many Egyptians as a vindication of Egyptian and Arab honor, and as something that the nation could glory in. What the Egyptians perceive as a victory has been memorialized in street names, public works projects, the names given to new settlements, and in stamps issued on the anniversary of the war's outbreak. Many of these stamps featured the Arab term for "crossing" (Al-ubur) – referencing the Egyptian crossing of the Suez Canal, which was seen as an important military and symbolic achievement.[51]

[50] See *The War in October* (Al Jazeera): part 2, first 25 minutes. For a detailed discussion of the role of the Iraqi army, see Andrew Terrill, "Iraq's Role in the 1973 Arab-Israeli War," *Small Wars & Insurgencies* 11(3) (2000).

[51] Yoram Meital, "Sadat's Grave and the Commemoration of the 1973 War in Egypt," in Michael Geisler (ed.), *National Symbols, Fractured Identities: Contesting the National Narrative* (Lebanon, NH: University Press of New England, 2005): pp. 225–26.

Sadat, in an expansive policy statement released in 1974, claimed that the war "released Arab potentials" and restored his people's dignity. From now on, the president argued, "[T]he Arab world is no longer booty over which the powerful quarrel, nor an area whose destiny is drawn up in remote capitals."[52] October, the month in which the war was launched, as well as the metaphor of crossing, became shorthand in Arabic for victory, for Egyptian prowess. Some years ago, the Israeli journalist Zvi Bar'el compared the commemoration of the war in both countries. He wrote:

> Compared to Israel, where you cannot find a street named after the Yom Kippur War, the Egyptian "October" has long since ceased to be the name of a month. October is a bridge, a school, a neighborhood, a weekly magazine, a sports club, a restaurant ... October is a symbol of the greatness of the Egyptian army which has become a model to be compared to any other colossal event in the country's history. Thus, for example, Egyptian publicists compared the revolution that overthrew Mubarak to the October victory.[53]

And here, more recently, is Egyptian President Abdel Fattah el-Sisi in a speech marking forty-six years to the October War: "Today, Egypt is celebrating one of its greatest and most glorious days, the day of crossing the line between despair and hope, the past and the future. The 6th of October 1973 does not only represent a significant military victory the Armed Forces have achieved aptly, but it is also a unique expression of a nation's will and people's unity. Egyptians were inspired by their great ancient history to remain strong and unbending ... It is the right of today's generations to know what happened in the war. We, who lived those days, have a duty to accurately clarify how the test was grueling, how Egyptians passed it successfully ... that the entire people became an army that pulled off out of that saga, a victory that reinstated dignity and opened the door for both land restoration and peace establishment ... I salute all the martyrs and heroes of the Armed Forces and the Egyptian people who sacrificed their lives for sacred Sinai. We tell their eternal souls on this day: 'Your children and grandchildren are willing to follow your suit and protect the free homeland you left under national sovereignty. Such homeland can never tolerate humiliation or submission; it does not get deceived by lies or defeated in psychological war. Egypt will always be a victorious homeland with a raised flag, and will always witness the defeats

[52] Sadat as quoted in Laurie Brand, *Official Stories: Politics and National Narratives in Egypt and Algeria* (Stanford, CA: Stanford University Press, 2014): p. 77.

[53] Zvi Bar'el, "In Egypt They are Still Fighting for Memory," *Haaretz* (September 13, 2013) (my translation).

and disappointments of its enemies no matter how the forms of war vary and how their mechanisms change.'"[54]

These excerpts testify to the relationship between humiliation and glory as we outlined in Chapter 7. Historical humiliation is such a potent, significant injury that its reversal must win super recognition. The reversal of humiliation becomes, to use Pericles' terms once again, the kind of national project, the kind of collective purpose, that is so important that those who take it up univariably win the "praise that never grows old." Terminating humiliation engenders glory.

8.4 Summary

The case studies discussed in this chapter illustrate the main theoretical claim of this book: Feelings of national humiliation fuel political fury; reversing these feelings, often by going to war against those perceived to have humiliated us is one of the most consequential national purposes a polity can undertake, and those who fight for that project will be awarded with the highest form of recognition on offer. Humiliation and glory are mutually reinforcing psychological forces that drive war forward. The rise of ISIS was animated by a narrative of the historical humiliation of Sunnis by "apostates." This narrative featured key elements of our account of humiliation in international affairs – from the dismissal of past promises to contempt towards cultural and geographical realities salient to the lives of the humiliated. This story of humiliation and the glory promised to those who would end it served as an effective tool for ISIS nation-building and recruitment. Russia's foreign policy in the last two decades is also deeply tied with a sense of national humiliation, both reflected and manufactured by Vladimir Putin, according to which Russia has been displaced and discarded as a serious world power by the United States and its NATO allies. Here, too, the phenomenology of humiliation as replacement is on display, as is the tendency to reward the reversal of humiliation with glory. Finally, we looked at the 1973 Middle East War as an example of a conflict fueled by an unbearable need to reverse an earlier humiliation. Egypt's primary aim in this war was erase or counteract the humiliation it suffered in the 1967 war with Israel. Interestingly, in this case, the officials who negotiated the war's conclusion took the sentiment's potency into

[54] Speech of President Abdel Fattah el-Sisi, October 6, 2019, available at the website of the Egyptian presidency: www.presidency.eg/en/الرئاسة/خطب-الرئيس/كلمة-الرئيس-عبد-الفتاح-السيسي-خلال-الاحتفال-بالذكرى-السادسة-والأربعين-لنصر-أكتوبر-المجيد/.

account as they designed the terms of the ceasefire and armistice. The historical examples we took up here also highlight the disconnect between the normative grounds for feeling humiliated and the psychological intensity of the feeling: Political entities can be overtaken by a sense of humiliation when they ought not to be (and, sometimes, don't feel humiliated when they have every right to be). The case studies in this chapter remind us that it is in our best interest to pay close attention to narratives of humiliation regardless of their normative status, as these narratives very often serve as "early warning" signs of impending violence.

Conclusion

Glory and humiliation drive war forward. They always have. From the Melian debate to the invasion of Ukraine, politicians, military commanders, and ordinary people refer to these terms to explain why they fight. As I conclude writing this book, the Middle East is, once more, engulfed by violence. Both sides to the war between Israel and Hamas deploy the language of humiliation and glory to make sense of what they are doing. In a recording released by the IDF, a member of Hamas' Nukhba force, which massacred scores of civilians in southern Israel on October 7, 2023, can be heard calling his commander and yelling into the phone, frenzied, that "the bastards are living happily while we are humiliated." He then vows to "take apart" every Israeli he sees.[1] The IDF's retaliatory campaign, during which thousands of civilians have been killed in Gaza, was framed, at least in part, as an effort to undo the humiliation the state suffered at the hands of Hamas. Four weeks into the war, the commander of the IDF's 36th Division sent out a communique to his subordinates in which he wrote: "[W]hat happened will not happen again . . . this is our war and today is our turn."[2] On the same day, Lieutenant Colonel Maoz Schwartz, the commander of a reserve battalion in the IDF's Jerusalem Brigade, told his soldiers, immediately before they entered the Gaza Strip: "[T]his is a historical hour. Only a few generations can say 'we went to fight for our homeland.' [We will] retrieve the honor that was lost on Simchat Tora and bring it back in spades [Simchat Tora is the Jewish holiday that October 7th fell on]." And then, as if echoing King Henry's promise of glory to his soldiers before the Battle of Agincourt in Shakespeare's play, Schwartz told his men that many years from now they will proudly tell their awed children and grandchildren how they fought Hamas after the massacre of 2023.[3]

[1] The recording and its Hebrew translation are available here: https://news.walla.co.il/item/3618836.

[2] Hebrew video available here (the translation is mine): www.makorrishon.co.il/news/688797/.

[3] Hebrew video available here (the translation is mine): www.inn.co.il/news/618348.

The focus of this book has been primarily descriptive: The theory offered in Chapter 5 provides resources for a normative assessment of claims for humiliation in international relations but that isn't my key aim here. Instead, I try to highlight the role that the dynamics of humiliation and glory, as a matter-of-fact, do play in political violence. As I've argued throughout, some of the humiliations that drive war are real, some manufactured. Sometimes those who report being humiliated don't have solid normative grounds for feeling the way they do. But these feelings do not become any less explosive for being groundless.

As we conclude, it's worth asking a question we have not really taken up: Is there a way of breaking out of the glory-humiliation cycle? The question rehashes Einstein's seemingly naïve query, with which we opened. Since the glory-humiliation dyad is central to fueling war, since it serves, as we have argued, as the two-stroke engine of conflict, asking whether its influence can be limited ends up pretty close to Einstein's inquiry about whether there is a way of "delivering mankind from the menace of war."[4]

One way of thinking about this is to consider whether people can become more suspicious about claims of national humiliation. Is there a way to convince them or teach them how susceptible that term is to abuse? Is there a way to show them how easy it is to manipulate polities into action in its name?

I'd like to think that any student of history would have learned this lesson. Anyone exposed to the history of manufacturing and inflating humiliations – from the German "knife in the back" to screams of replacement in Charlottesville in 2017 – should develop a baseline suspicion of the term. And yet, in the first place, there are not as many students of history as we would like there to be. And, even those who are well-versed in it, are far better at spotting old stupidities than at identifying the echoes of past mistakes in their own conduct. Thus, for example, the planners of the Iraq war must have known how frustrating and degrading it was for Vietnam veterans to be sent into a war they could not win. They had to have known about the sense of betrayal and humiliation this created among returning soldiers.[5] And yet, mindlessly, they sent a brand-new generation of soldiers into another war with an ill-defined end game and expected different results.

4 www.public.asu.edu/~jmlynch/273/documents/FreudEinstein.pdf.

5 For an outstanding account of these reactions, see Jonathan Shay, *Achilles in Vietnam: Combat Trauma and the Undoing of Character* (New York: Scribner, 1994): chapter 1.

If not in the study of history, perhaps hope is to be found in what James called the "moral equivalent of war"?[6] Could we learn to become excited about more pacific causes than the martial reversal of national humiliation? Is there a way to convince people to glorify something else except fighting to end these indignities? The remarkable staying power of the glory-humiliation dynamic through the millennia provides a prima facie case against James' optimism. Then again perhaps we have not tried enough. In this book, I have focused on understanding how these two great moral passions move war forward. What it would take to find a moral equivalent of war, whether that idea can really be philosophically defended, whether there are environmental or social or scientific causes that can sufficiently excite people while also sating their need for honor and glory are aims for an entirely different (and more hopeful) book. But we might note that finding these causes and attaching people to them would be a political education project that would take a very long time. So long, perhaps, that one might be forgiven for once again conjuring up Freud's stark image of "mills that grind so slowly that, before the flour is ready, men are dead of hunger."[7]

If it doesn't provide a remedy to the problem it diagnosis, at the very least I hope this book will direct attention to the weight of these deadly dynamics: knowing how they play out, knowing how psychologically powerful they are even when they are divorced from moral entitlements, can serve as an important early warning sign for impending violence. Whatever they thought about the benefits of globalization, officials in the United States should have become much more concerned, much earlier, about the sense of humiliation and despair that ran rampant in the country's rust belt in the last twenty years. Whatever they thought of Boris Yeltsin or Vladimir Putin, however relieved they were about the end of the Cold War, however exuberant they were about the advent of the liberal international order, successive American presidents and secretaries of state should have known that there was a real risk in the dismissive treatment of the Russians beginning in the early 1990s. They should have anticipated that snubbing them would play into the hands of those intent on building a new Russian nationalism. The normative questions about whether we really have humiliated others, and the attempt to figure out how much of the humiliation they feel is really our fault, are important and difficult to answer. But in the end, these are not exclusively, perhaps even not primarily, moral matters. Almost invariably, a palpable sense of political humiliation, whoever is to blame for it, is a reliable sign of bad things to come. Taking that seriously is the very essence of political prudence.

[6] James, "Moral Equivalent of War."

[7] See the Introduction to this book.

Bibliography

Ajami, Fouad, *The Arab Predicament: Arab Political Thought and Practice since 1967* (Cambridge, UK: Cambridge University Press, 1992).

Alexander, Caroline, *The War that Killed Achilles* (New York: Penguin, 2010).

Allison, Graham, *Destined for War: Can America and China Escape Thucydides's Trap?* (Boston, MA: Houghton Mifflin Harcourt, 2017).

Ambrose, Stephen, *Citizen Soldiers: The U. S. Army from the Normandy Beaches to the Bulge to the Surrender of Germany, June 7, 1944–May 7, 1945* (New York: Simon and Schuster, 1997).

Amery, Jean, "Torture" in *At the Mind's Limits* trans. Rosenfeld and Rosenfeld (Bloomington, IN: Indiana University Press, 2009), pp. 21–41.

Anderson, Eric, *China Restored: The Middle Kingdom Looks to 2020 and Beyond* (Santa Barbara, CA: Praeger, 2010).

Appiah, Anthony, *The Honor Code* (New York: Norton, 2010).

Arendt, Hannah, "For the Honor and Glory of the Jewish People" in *The Jewish Writings* eds. Kohn and Feldman (New York: Schocken Books, 2007): 199–200.

The Human Condition (Chicago, IL: The University of Chicago Press, 1989).

Aristotle, *The Politics* trans. Saunders (New York: Penguin, 1981).

Auden, Wystan Hugh, *Collected Poems* (New York: Vintage, 1991).

Badie, Bertrand, *Humiliation in International Relations: A Pathology of Contemporary International Systems* (Portland, OR: Hart Publishing, 2017).

Bar-Joseph, Uri, "Lessons not Learned: Israel in the Post-Yom Kippur War Era," *Israel Affairs*, 14(1) (2008), 70–83.

Bar'el, Zvi, "In Egypt They are Still Fighting for Memory," *Haaretz* (September 13, 2013).

Barnhart, Joslyn, *The Consequences of Humiliation: Anger and Status in World Politics* (Ithaca, NY: Cornell University Press, 2020).

Barr, James, *A Line in the Sand: Britain, France and the Struggle for the Mastery of the Middle East* (New York: Simon and Schuster, 2011).

Bass, Gary, *Stay the Hand of Vengeance: The Politics of War Crimes Tribunals* (Princeton, NJ: Princeton University Press, 2002).

Berger, Jonah and Milkman, Katherine, "What Makes Online Content Viral?" *Journal of Marketing Research*, 49(2) (2012), 192–205.

Bird, Kai and Sherwin, Martin, *American Prometheus: The Triumphs and Tragedy of J. Robert Oppenheimer* (New York: Vintage Books, 2006).

Birmingham, Peg, "Arendt and Hobbes: Glory, Sacrificial Violence, and the Political Imagination," *Research in Phenomenology*, 41(1) (2011), 1–22.

Bobo, Lo, *Russia and the New World Disorder* (London, UK: Brookings Institution Press, 2015).

Boswell, James, *Life of Johnson* ed. Chapman (New York: Oxford University Press, 1980).

Bouverie, Tim, *Appeasing Hitler: Chamberlain, Churchill, and the Road to War* (London, UK: Vintage, 2019).

Brand, Laurie, *Official Stories: Politics and National Narratives in Egypt and Algeria* (Stanford, CA: Stanford University Press, 2014).

Brickhouse, Thomas and Smith, Nicholas, "Socrates on How Wrongdoing Damages the Soul," *The Journal of Ethics*, 11(4) (2007), 337–56.

Burke, Edmund, *Reflections on the Revolution in France* (Indianapolis, IN: Hackett, 1987).

Burley, Jake and Eisikovits, Nir, "Workplace Automation and Political Replacement: A Valid Analogy?" *AI and Ethics*, 3(4) (2022), 1361–70.

Camus, Renaud, *Le Grand Remplacement* (Paris, France: Chez l'auteur, 2011).

Carpenter, Ted Galen, "Many Predicted NATO Expansion would Lead to War. Those Warnings were Ignored," *The Guardian* (February 28, 2002). www.theguardian.com/commentisfree/2022/feb/28/nato-expansion-war-russia-ukraine.

Cassel, Christopher, *To End All War: Oppenheimer & the Atomic Bomb* (NBC/Universal Documentary, 2023).

Chappell, Timothy, "Glory as an Ethical Idea," *Philosophical Investigations*, 34(2) (2011), 105–34.

Chua, Amy, "Tribal World: Group Identity is All," *Foreign Affairs* (July/August 2018).

Cicero, *On Obligations* trans. Walsh (New York: Oxford University Press, 2008).

Clausewitz, Carl von, *On War* trans. Howard and Paret (Princeton, NJ: Princeton University Press, 1984).

Crozier, Ray, "Differentiating Shame from Embarrassment," *Emotion Review*, 6 (3) (2014), 269–76.

Dao, James, "Drone Pilots are Found to Get Stress Disorders Much as Those in Combat Do," *The New York Times* (February 22, 2013). www.nytimes.com/2013/02/23/us/drone-pilots-found-to-get-stress-disorders-much-as-those-in-combat-do.html.

Day-Lewis, Cecil (ed.), *The Collected Poems of Wilfred Owen* (New York: New Directions Books, 1963).

Demetriou, Dan, "Honour" in *Routledge Encyclopedia of Philosophy* (Taylor and Francis, 2020). www.rep.routledge.com/articles/thematic/honour/v-2. https://doi.org/10.4324/9780415249126-L036-2.

Deonna, Julien, Rodogno, Raffaele and Teroni, Fabrice, *In Defense of Shame: The Faces of an Emotion* (Oxford, UK: Oxford University Press, 2013).

Dudai, Ron and Cohen, Hillel, "Triangle of Betrayal: Collaborators and Transitional Justice in the Israeli-Palestinian Conflict," *Journal of Human Rights*, 6(1) (2007), 37–58. https://doi.org/10.1080/14754830601084642.

Dutmer, Evan, "Summa et Perfecta Gloria: Cicero on Ambition, Reputation, and Care for Future Human Beings," *Ethical Perspectives*, 29(1) (2022), 7–31.

Eisikovits, Nir, *Sympathizing with the Enemy: Reconciliation, Transitional Justice, Negotiation* (Leiden, The Netherlands: Brill/Martinus Nijhoff, 2010).

"Proportionality and Self-Interest," *Human Rights Review*, 11(2) (2010), 157–70.

"Peace versus Justice in Transitional Settings," *Quinnipiac Law Review*, 32(3) (2014), 221–36.

A Theory of Truces (London, UK: Palgrave MacMillan, 2015).

"Transitional Justice" in *The Stanford Encyclopedia of Philosophy* ed. Zalta (Fall 2017 Edition). https://plato.stanford.edu/archives/fall2017/entries/justice-transitional.

"Embarrassment and Political Repair" in *Emotions and Mass Atrocity: Philosophical and Theoretical Explorations* eds. Brudholm and Lang (Cambridge, UK: Cambridge University Press, 2018), pp. 262–76.

"Not Set in Stone: Five Bad Arguments for Letting Monuments Stand," *Journal of Global Ethics*, 16(3) (2020), 404–13.

"Political Humiliation and the Sense of Replacement" in *How to End a War* eds. Parsons and Wilson (New York: Cambridge University Press, 2023).

Fanon, Frantz, *The Wretched of the Earth* (New York: Grove Press, 2004).

Fawcett, Edmund, *Conservatism: The Fight for a Tradition* (Princeton, NJ: Princeton University Press, 2020).

French, Shannon, *The Code of the Warrior: Exploring Warrior Values Past and Present* (Lanham, MD: Rowman & Littlefield, 2003).

Freud, Sigmund, *Civilization and its Discontents* trans. Strachey (New York: Norton, 1961).

Frevert, Ute, *The Politics of Humiliation* (New York: Oxford University Press, 2020).

Fukuyama, Francis, *Identity: The Demand for Dignity and the Politics of Resentment* (New York: Farrar, Straus and Giroux, 2018).

Fussell, Paul, *The Great War and Modern Memory* (New York: Oxford University Press, 2000).

Fussi, Alessandra, "Aristotle on Shame," *Ancient Philosophy*, 35(1) (2015), 113–35.

Gaddis, John Lewis, *The Cold War: A New History* (London, UK: Penguin Books, 2005).

Gan, Alon, "'Till Death,' 'Brothers Sitting in the Shadow of Cannons': Oz and Levin as Anti-Euphoric Prophets" in *Israel 67–77 Continuity and Turning Point* eds. Shif and Halamish (Beer Sheva, Israel: Ben-Gurion Institute, 2017), p. 338 (Hebrew, my translation).

Gerges, Fawaz, *ISIS: A History* (Princeton, NJ: Princeton University Press, 2017).

Gilbert, Mark, "From War to Cold War" in *A Companion to Europe since 1945* ed. Larres (Hoboken, NJ: Blackwell, 2009).

Gilbert, Paul and Gerlsma, Coby, "Recall of Shame and Favouritism in Relation to Psychopathology," *British Journal of Clinical Psychology*, 38(4) (1999), 357–73.

Gorenberg, Gershom, *The Accidental Empire: Israel and the Birth of the Settlements, 1967–1977* (New York: Holt, 2007).

Gornick, Vivian, "Put on the Diamonds," *Harper's* (October 4, 2021). https://harpers.org/archive/2021/10/put-on-the-diamonds-notes-on-humiliation-vivian-gornick/.

Greenberg, Kenneth, *Honor and Slavery* (Princeton, NJ: Princeton University Press, 1997).

Griswold, Charles, *Forgiveness: A Philosophical Exploration* (New York: Cambridge University Press, 2007).

Gutmann, Amy and Thompson, Dennis, "The Moral Foundations of Truth Commissions" in *Truth v. Justice: The Morality of Truth Commissions* eds. Rotberg and Thompson (Princeton, NJ: Princeton University Press, 2000).

Hamner, Christopher, *Enduring Battle: American Soldiers in Three Wars, 1776-1945* (Lawrence, KS: University Press of Kansas, 2011).

Harris, Arthur T., *Dispatch on War Operations: 23rd February 1942 to 8th May 1945* (London, UK: Frank Cass, 1995).

Hart, Herbert, *The Concept of Law* (New York: Oxford University Press, 1994).

Hastings, Mark, Northman, Lisa and Tangney, June, "Shame, Guilt, and Suicide" in *Suicide Science: Expanding the Boundaries* eds. Joiner and Rudd (Boston, MA: Springer, 2002).

Hemingway, Ernest, *A Farewell to Arms* (New York: Scribner, 2014).

Hirschman, Albert, *The Passions and the Interests: Political Arguments for Capitalism before its Triumph* (Princeton, NJ: Princeton University Press, 2013).

Hitchens, Christopher, "Edmund Burke: Reactionary Prophet," *The Atlantic* (April 2004 Issue). www.theatlantic.com/magazine/archive/2004/04/reactionary-prophet/302914/.

Hobbes, Thomas, *Leviathan* ed. and intro. Curley (Indianapolis, IN: Hackett, 1994).

Homer, *Iliad* trans. Fagles (New York: Penguin, 1988).

The Odyssey trans. Fagles (New York: Penguin, 1997).

Honneth, Axel, "Integrity and Disrespect: Principles of a Conception of Morality Based on the Theory of Recognition," *Political Theory*, 20(2) (1992), 187–201.

Ignatieff, Michael, "Digging up the Dead," *The New Yorker* (November 10, 1997).

Ingram, Haroro, Whiteside, Craig and Winter, Charlie, *The ISIS Reader: Milestone Texts of the Islamic State Movement* (New York: Oxford University Press, 2020).

James, William, "The Moral Equivalent of War," *Lecture 11 in Memories and Studies* (New York: Longmans, Green & Co, 1911), pp. 267–96.

Jünger, Ernst, *In Stahlgewittern: Aus dem Tagebuch eines Stosstruppführers* (Leipzig, Germany: Robert Meiser, 1920), translated into English as *The Storm of Steel* (London, UK: Chatto and Windus, 1929).

Storm of Steel trans. Hoffman (New York: Penguin, 2004).

Kagan, Donald, *The Peloponnesian War* (New York: Penguin, 2004).

Kant, Immanuel, "Perpetual Peace" in *The Global Justice Reader* ed. Brooks (New York: Wiley, 2008).

Kaufman, Uri, *Eighteen Days in October: The Yom Kippur War and How It Created the Modern Middle East* (New York: St Martin's Press, 2023).

Keynes, John Maynard, *The Economic Consequences of the Peace* (London, UK: Empire Books, 2012).

Khorram-Manesh, Amir, Burkle, Frederick, Goniewicz, Krzysztof and Robinson Yohan, "Estimating the Number of Civilian Casualties in Modern Armed Conflicts–A Systematic Review," *Frontiers in Public Health*, 9 (2021). https://doi.org/10.3389/fpubh.2021.765261.

Kimerling, Baruch, *The Palestinian People: A History* (Cambridge, MA: Harvard University Press, 2003).

Kipling, Rudyard, "Harp Song of the Dane Women." www.poetryfoundation .org/poems/47691/harp-song-of-the-dane-women.

Kiss, Elizabeth, "Moral Ambition within and beyond Political Constraints: Reflections on Restorative Justice" in *Truth v. Justice: The Morality of Truth Commissions* eds. Rotberg and Thompson (Princeton, NJ: Princeton University Press, 2000).

Kissinger, Henry, *Years of Upheaval* (Boston, MA: Little, Brown, 1982).

Krems, Jaimie, Williams, Keelah, Aktipis, C. Athena and Kenrick, Douglas T., "Friendship Jealousy: One Tool for Maintaining Friendships in the Face of Third-Party Threats?" *Journal of Personality and Social Psychology*, 120(4) (2020), 977–1012.

Langworth, Richard, *Churchill in His Own Words* (London, UK: Ebury Press, 2012).

Leask, Phil, "Losing Trust in the World: Humiliation and its Consequences," *Psychodynamic Practice*, 19(2) (2013), 129–42.

Leconte, Cecile, "The Socio-Political Career of the Expression "The Great Replacement" among Right-Wing Party Networks in Germany: The Case of the Alternative for Germany (AfD) party," *Politix*, 126(2) (2019), 111–34.

Lindner, Evelin, *Making Enemies: Humiliation and International Conflict* (Westport, CT: Praeger, 2006).

Honor, Humiliation, and Terror: An Explosive Mix – And How We Can Diffuse It with Dignity (Dörzbach, Germany: Dignity Press, 2017).

Long, Anthony, "Cicero's Politics in *De Officiis*" in *From Epicurus to Epictetus: Studies in Hellenistic and Roman Philosophy* ed. Long (Oxford, UK: Oxford University Press, 2006).

Lu, Catherine, "Shame, Guilt and Reconciliation after War," *European Journal of Social Theory*, 11(3) (2008), 367–83.

Maatouk, Tamara, "On the Heels of 1967: Chahine, Cinema, and Emotional Response(s) to the Defeat," *International Journal of Middle East Studies*, 55 (1) (2023), 25–42.

Machiavelli, Niccolo, *The Prince* trans. Alvarez (Long Grove, IL: Waveland Press, 1989).
Discourses on Livy trans. Bondanella and Bondanella (New York: Oxford University Press, 2008).
Mandeville, Bernard, *The Fable of the Bees or Private Vices, Public Benefits* (London, UK: T. Ostell, 1806 – via the Gutenberg Project). www.gutenberg.org/cache/epub/57260/pg57260-images.html.
Margalit, Avishai, *The Decent Society* (Cambridge, MA: Harvard University Press, 1996).
Mark, Clifton, "The Natural Laws of Good Manners: Hobbes, Glory, and Early Modern Civility," *The Review of Politics,* 80(3) (2018), 391–414.
McCormick, John, "Machiavelli's Inglorious Tyrants: On Agathocles, Scipio and Unmerited Glory," *History of Political Thought* 36(1) (2015), 29–52.
Mchangama, Jacob and Verdirame, Guglielmo, "The Danger of Human Rights Proliferation: When Defending Liberty, Less Is More," *Foreign Affairs* (July 2013).
McPherson, James, *For Cause and Comrades: Why Men Fought in the Civil War* (New York: Oxford University Press, 1997).
Mearsheimer, John, "Why the Ukraine Crisis is the West's Fault: The Liberal Delusions that Provoked Putin," *Foreign Affairs* (September/October 2014).
Meital, Yoram, "Sadat's Grave and the Commemoration of the 1973 War in Egypt" in *National Symbols, Fractured Identities: Contesting the National Narrative* ed. Geisler (Lebanon, NH: University Press of New England, 2005).
Melhem, Hisham, "The Arab World has Never Recovered from the Loss of 1967," *Foreign Policy* (June 5, 2017). https://foreignpolicy.com/2017/06/05/the-arab-world-has-never-recovered-from-the-loss-of-1967/.
Menon, Rajan, "How to End the War in Ukraine," *Boston Review* (April 26, 2023). www.bostonreview.net/forum/how-to-end-the-war-in-ukraine/.
Menon, Shivshankar, "How China Bucked Western Expectations and What it Means for World Order," *Brookings Order From Chaos Blog* (2016). www.brookings.edu/blog/order-from-chaos/2016/03/10/how-china-bucked-western-expectations-and-what-it-means-for-world-order/.
Miller, Reuben, "The Israeli-Syrian Negotiations," *Mediterranean Quarterly,* 11 (4) (2000), 117–39.
Moggridge, Donald (ed.), *The Collected Writings of John Maynard Keynes Volume CWXXVIII: Political and Literary Writings* (London, UK: Macmillan for the Royal Economic Society, 1982).
Montefiore, Simon, *Stalin: The Court of the Red Tsar* (New York: Vintage, 2005).
Müller-Markus, Christina, "One Belt, One Road: The Chinese Dream and its Impact on Europe," *Notes Internacionals CIDOB*, 148, (2016).
Murdoch, Simon and Mulhall, Joe, "From Banners to Bullets: The International Identarian Movement" (London, UK: HOPE not hate Charitable Trust, 2019). www.hopenothate.org.uk/wp-content/uploads/2019/12/Identitarianism-Report.pdf.

Murphy, Colleen, *A Moral Theory of Political Reconciliation* (New York: Cambridge University Press, 2010).
The Conceptual Foundations of Transitional Justice (New York: Cambridge University Press, 2017).
Nagel, Thomas, "Moral Luck" in *Moral Luck* ed. Statman (Albany, NY: SUNY Press, 1993).
Nussbaum, Martha, "Danger to Human Dignity: The Revival of Disgust and Shame in the Law," *The Chronicle of Higher Education* (August 6, 2004).
Hiding from Humanity: Disgust, Shame, and the Law (Princeton, NJ: Princeton University Press, 2004).
Oren, Michael, *Six Days of War: June 1967 and the Making of the Modern Middle East* (New York: Presidio, 2003).
Orwell, George, "Review of Mein Kampf" in *George Orwell: Essays* (New York: Everyman's Library, 2002), pp. 249–50.
Owen, David, "Machiavelli's *il Principe* and the Politics of Glory," *European Journal of Political Theory*, 16(1) (2015), 41–60.
Palmer, Michael, "Love of Glory and the Common Good," *American Political Science Review*, 76(4) (1982), 825–36.
Pape, Robert, *Dying to Win: The Strategic Logic of Suicide Terrorism* (New York: Random House, 2005).
Paul, T. V., "Taboo or Tradition? The Non-Use of Nuclear Weapons in World Politics," *Review of International Studies, 36*(4) (2010), 853–63.
Petersen, Roger, "Guilt, Shame, Balts, Jews" in *Confronting Memories of World War II: European and Asian Legacies* eds. Chirot, Shin and Sneider (Seattle, WA: University of Washington Press, 2014), pp. 258–83.
Phelps, Hattie, "The Disproportionate Effects of War and Conflict on Women and Girls," *The Georgetown Journal of Gender and the Law*, 23(3) (2003).
Philips, Adam, "The Magical Act of a Desperate Person," *London Review of Books*, 35(5) (2013).
Pontecorvo, Gillo, *The Battle of Algiers* (Igor Film and Casbah Film, 1966).
Pope, Sterett, "The Education of Hafez al-Assad," *Worldview*, 28(4) (1985), 4–8.
Press, Eyal, "The Wounds of the Drone Warrior," *The New York Times* (June 13, 2018). www.nytimes.com/2018/06/13/magazine/veterans-ptsd-drone-warrior-wounds.html.
Price, Russell, "The Theme of Gloria in Machiavelli," *Renaissance Quarterly*, 30(4) (1977), 588–631.
Putin, Vladimir, "Russia at the Turn of the Millennium," *Nezavisimaya Gazeta* (December 30, 1999) (Russian translation courtesy of Alan Zebek).
Rabinovich, Abraham, *The Yom Kippur War: The Epic Encounter that Transformed the Middle East* (New York: Schocken Books, 2017).
Rabinovich, Oded and Stein, Kenneth, "The October 1973 War: Superpower Engagement and Estrangement," (Hebrew) *Zmanim* 84 (2003), 59–69.
Reid, Frances and Hoffmann, Deborah, *Long Night's Journey into Day* (Iris Films, 2001).

Renan, Ernest, "What is a Nation?" Text of a Lecture Delivered at the Sorbonne on March 11th, 1882 in *Qu'est-ce qu'une nation?* trans. Rundell (Paris, France: Presses Pocket, 1992). http://ucparis.fr/files/9313/6549/9943/What_is_a_Nation.pdf.

Reynolds, David, *In Command of History: Churchill Fighting and Writing the Second World War* (London, UK: Allen Lane, 2004).

Rolnik, Eran, "When Freud and Einstein Tried to Answer an Age-old Question: Why War?" *Haaretz* (May 25, 2017). www.haaretz.com/life/books/2017-05-25/ty-article-magazine/.premium/when-freud-and-einstein-tried-to-answer-an-age-old-question-why-war/0000017f-f755-d044-adff-f7fda5230000#.

Rose, Alexander, *Washington's Spies: The Story of America's First Spy Ring* (New York: Bantam, 2014).

Rostron, Bryan, *Robert McBride: The Struggle Continues* (Cape Town, South Africa: Tafelberg, 2019).

Rotberg, Robert and Thompson, Dennis (eds.), *Truth v. Justice: The Morality of Truth Commissions* (Princeton, NJ: Princeton University Press, 2000).

Sadat, Anwar, *In Search of Identity: An Autobiography* (New York: HarperCollins, 1978).

Sánchez, Alba Montes and Zahavi, Dan, "Unraveling the Meaning of Survivor Shame" in *Emotions and Mass Atrocity: Philosophical and Theoretical Explorations* eds. Brudholm and Lang (Cambridge, UK: Cambridge University Press, 2018).

Saurette, Paul, *The Kantian Imperative: Humiliation, Common Sense, Politics* (Toronto, CA: University of Toronto Press, 2005).

Segal, Charles, "Kleos and its Ironies in the Odyssey," *L'antiquité Classique*, 52 (1983), 22–47.

Shakespeare, William, *Henry V* (New York: Penguin, 2015).

Shay, Jonathan, *Achilles in Vietnam: Combat Trauma and the Undoing of Character* (New York: Scribner, 1994).

Shelley, Mary, *Frankenstein* (London, UK: Collectors Library, 2017).

Sherman, Nancy, *Stoic Warriors: The Ancient Philosophy behind the Military Mind* (New York: Oxford University Press, 2007).

Afterwar: Healing the Moral Wounds of Our Soldiers (New York: Oxford University Press, 2015).

Singer, Peter, "Famine, Affluence, and Morality," *Philosophy and Public Affairs*, 1 (3) (1972), 229–43.

Singer, Peter and Brooking, Emerson, *LikeWar: The Weaponization of Social Media* (New York: Eamon Dolan, 2018).

Skinner, Quentin, *Machiavelli: A Very Short Introduction* (New York: Oxford University Press, 2001).

Visions of Politics Vol 2: Renaissance Virtues (Cambridge, UK: Cambridge University Press, 2002).

Slavicek, Louise, *The Treaty of Versailles. Milestones in Modern World History* (New York: Chelsea House Publications, 2010).

Slomp, Gabriella, "From 'Genus' to 'Species': The Unravelling of Hobbesian Glory," *History of Political Thought*, 19(4) (1998), 552–69.

Thomas Hobbes and the Political Philosophy of Glory (London, UK: Palgrave MacMillan, 2000).

Smith, Brian, "Two Dogmas of the New War Thesis," *International Journal of Military History and Historiography*, 38(1) (2018), 92–120.

Sophocles, *Ajax* trans. Jameson (Chicago, IL: The University of Chicago Press, 2013).

Sophocles, *Ajax* trans. Moore in *Sophocles II* eds. Grene and Lattimore (Chicago, IL: The University of Chicago Press, 2013).

Specter, Michael, "Letter from Moscow: Why are Vladimir Putin's Enemies Dying?" *The New Yorker* (January 29, 2007).

Srinivasa, Murthy and Lakshminarayana, Rashmi, "Mental Health Consequences of War: A Brief Review of Research Findings," *World Psychiatry: Official Journal of the World Psychiatric Association,* 5(1) (2006), 25–30.

Stewart, Frank, *Honor* (Chicago, IL: The University of Chicago Press, 1994).

Stewart, Rory and Knaus, Gerald, *Can Intervention Work?* (New York: Norton, 2012).

Suny, Ronald, "Ukraine War Follows Decades of Warnings that NATO Expansion into Eastern Europe could Provoke Russia," *The Conversation* (February 28, 2022). https://theconversation.com/ukraine-war-follows-decades-of-warnings-that-nato-expansion-into-eastern-europe-could-provoke-russia-177999.

Tal, David, "Who Needed the October 1973 War?" *Middle Eastern Studies,* 52(5) (2016), 748–49.

Terman, David, "Paranoid Leadership" in *The Leader* eds. Strozier, Offer and Abdyli (New York: Springer, 2011).

Terrill, Andrew, "Iraq's Role in the 1973 Arab-Israeli War," *Small Wars & Insurgencies*, 11(3) (2000).

The War in October (Al Jazeera Documentary, 2013).

Thucydides, *On Justice, Power and Human Nature: Selections from the History of the Peloponnesian War* trans. Woodruff (Indianapolis, IN: Hackett, 1993).

Tooby, John and Cosmides, Leda, "Friendship and the Banker's Paradox: Other Pathways to the Evolution of Adaptations for Altruism" in *Evolution of Social Behaviour Patterns in Primates and Man* eds. Runciman, Smith and Dunbar, Proceedings of The British Academy 88 (New York: Oxford University Press, 1996), pp. 119–43.

Tracy, Jessica and Robins, Richard, "Putting the Self into Self-Conscious Emotions: A Theoretical Model," *Psychological Inquiry*, 15(2) (2004), 103–25.

Tripathi, Deepak, *Imperial Designs: War, Humiliation & the Making of History* (Washington, D.C.: Potomac Books, 2013).

Tuck, Richard, *Hobbes: A Very Short Introduction* (New York: Oxford University Press, 1989).

Tutu, Desmond, *No Future without Forgiveness* (New York: Image/Doubleday, 2000).

Van Creveld, Martin, *The Transformation of War* (New York: Free Press, 1991).

von Kleist, Heinrich, *Michael Kohlhaas* trans. Greenberg (Hoboken, NJ: Melville House Publishing, 2005).

Waldron, Jeremy, "Superseding Historical Injustice," *Ethics*, 103(1) (1992), 4–28.

Walker, Shaun, "The Humiliation that Pushed Putin to Try and Recapture Russian Glory," *History* (February 23, 2018). www.history.com/news/vladimir-putin-russia-power.

The Long Hangover: Putin's New Russia and the Ghosts of the Past (New York: Oxford University Press, 2019).

Walzer, Michael, *Just and Unjust Wars* (New York: Basic Books, 2015).

Wheatcroft, Geoffrey, *Churchill's Shadow: The Life and Afterlife of Winston Churchill* (New York: Norton, 2021).

Whitehouse, Harvey, McQuinn, Brian, Buhrmester, Michael and Swann Jr., William B., "Brothers in Arms: Libyan Revolutionaries Bond Like Family," *Proceedings of the National Academy of Sciences of the United States of America*, 111(50) (2014), 17783–85.

Williams, Bernard, "Moral Luck" in *Moral Luck* ed. Statman (Albany, NY: SUNY Press, 1993).

Yarovskaya, Marianna (producer), *How to Become a Tyrant*, Netflix Documentary Series (Estuary Films, 2021).

Yu, Hong, "Motivation behind China's 'One Belt, One Road' Initiatives and Establishment of the Asian Infrastructure Investment Bank," *Journal of Contemporary China*, 26(105) (2017), 353–68.

Zabad, Ibrahim, "Egypt: Defeat and the Transformation of State and Society" in *The Aftermath of Defeats in War* (London, UK: Palgrave Macmillan, 2019).

Zheng, Wang, *Never Forget National Humiliation: Historical Memory in Chinese Politics and Foreign Relations* (New York: Columbia University Press, 2014).

Zicheng, Ye, *Inside China's Grand Strategy: The Perspective from the Peoples' Republic* trans. Levine and Liu (Lexington, KY: The University Press of Kentucky, 2011).

Index